Fractals and chaos are relevant to the understanding of many earth processes. This book introduces the fundamental concepts of fractal geometry and chaotic dynamics. It then relates them to a variety of geological and geophysical problems and illustrates what chaos theory and fractals can really tell us and how they can be applied to the earth sciences. The concepts are introduced at the lowest possible level of mathematics consistent with their understanding, such that the reader requires only a background in basic physics and mathematics. Petroleum and mineral reserves, earthquakes, mantle convection, and magnetic field generation are among the earth's properties that come under scrutiny.

Fractals and chaos in geology and geophysics can be used as a text for advanced undergraduate and introductory graduate courses in the physical sciences. Problems are included for the reader to solve.

Fractals and chaos in geology and geophysics

Fractals and chaos in geology and geophysics

DONALD L. TURCOTTE

Maxwell Upson Professor of Engineering
Department of Geological Sciences
Cornell University, Ithaca, New York

CAMBRIDGE
UNIVERSITY PRESS

Published by the Press Syndicate of the University of Cambridge
The Pitt Building, Trumpington Street, Cambridge CB2 1RP
40 West 20th Street, New York, NY 10011-4211, USA
10 Stamford Road, Oakleigh, Victoria 3166, Australia

First published 1992

Printed in Great Britain at the University Press, Cambridge

A catalogue record for this book is available from the British Library

Library of Congress cataloguing in publication data

Turcotte, Donald Lawson.
    Fractals and chaos in geology and geophysics/Donald L. Turcotte.
        p.   cm.
    Includes bibliographical references and index.
    ISBN 0 521 41270 6
    1. Geology–Mathematics.   2. Geophysics–Mathematics.
    3. Fractals.   4. Chaotic behavior in systems.   I. Title.
    QE33.2.M3T87   1992
    550′.1′51474–dc20   91-27970 CIP

    ISBN 0 521 41270 6 hardback

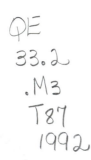

Contents

	Preface	ix
1	Scale invariance	1
2	Definition of a fractal set	6
3	Fragmentation	20
4	Seismicity and tectonics	35
5	Ore grade and tonnage	52
6	Fractal clustering	65
7	Self-affine fractals	73
8	Geomorphology	95
9	Dynamical systems	104
10	Logistic map	114
11	Slider-block models	125
12	Lorenz equations	137
13	Is mantle convection chaotic?	151
14	Rikitake dynamo	162
15	Renormalization group method	169
16	Self-organized criticality	186
17	Where do we stand?	195
	References	196
	Appendix A	206
	Appendix B	209
	Answers to selected problems	213
	Index	216

Preface

I was introduced to the world of fractals and renormalization groups by Bob Smalley in 1981. At that time Bob had transferred from physics to geology at Cornell as a Ph.D. student. He organized a series of seminars and convinced me of the relevance of these techniques to geological and geophysical problems. Although his official Ph.D. research was in observational seismology, Bob completed several renormalization and fractal projects with me. Subsequently, my graduate students Jie Huang and Cheryl Stewart have greatly broadened my views of the world of chaos and dynamical systems. Original research carried out by these students is included throughout this book.

The purpose of this book is to introduce the fundamental principles of fractals, chaos, and aspects of dynamical systems in the context of geological and geophysical problems. My goal is to introduce the fundamental concepts at the lowest level of mathematics that is consistent with the understanding and application of the concepts. It is clearly impossible to discuss all aspects of applications. I have tried to make the applications reasonably comprehensible to non-earth scientists but may not have succeeded in all cases.

After an introduction, the next seven chapters are devoted to fractals. The fundamental concepts of self-similar fractals are introduced in Chapter 2. Applications of self-similar fractals to fragmentation, seismicity and tectonics, ore grades and tonnage, and clustering are given in the next chapters. Self-affine fractals are introduced in Chapter 7 and are applied to geomorphology in Chapter 8. A brief introduction to dynamical systems is given in Chapter 9. The fundamental concepts of chaos are introduced through the logistic map, slider-block models, the Lorenz equations, mantle convection, and the Rikitake dynamo in the next five chapters. The renormalization group method is introduced in Chapter 15 and self-organized criticality is considered in Chapter 16.

Problems are included so that this book can be used as a higher-level undergraduate text or a graduate text, depending upon the background of the students and the material used. Little mathematical background is required for the introduction to self-similar fractals and chaos that includes Chapters 1–6 and 10–11. The treatment of self-affine fractals in Chapters 7–8 requires some knowledge of spectral techniques. Chapters 9 and 12–13 require a knowledge of differential equations.

I would like to dedicate this book to the memory of Ted Flynn. Until his untimely death in 1989 Ted was Chief of the Geodynamics Branch at NASA Headquarters. In this position, over a period of ten years, he supervised a program that changed routine centimeter-level geodetic position measurements from a dream to a reality. Ted also had the foresight to support research on fractals and chaos applied to crustal dynamics at a time when these subjects were anything but popular. In particular his enthusiasm was instrumental in a conference on earthquakes, fractals, and chaos held at the Asilomar conference facility in January 1989. This conference established a dialogue between physicists, applied mathematicians, and seismologists focused on the applications of dynamical systems to earthquake prediction.

I would also like to acknowledge extensive discussions with John Rundle, Charlie Sammis, Chris Barton, and Per Bak. Chapter 8 is largely the result of a collaboration with Bill Newman. And this book could not have been completed without the diligent manuscript preparation of Maria Petricola.

CHAPTER ONE

Scale invariance

A stone, when it is examined, will be found a mountain in miniature. The fineness of Nature's work is so great, that, into a single block, a foot or two in diameter, she can compress as many changes of form and structure, on a small scale, as she needs for her mountains on a large one; and, taking moss for forests, and grains of crystal for crags, the surface of a stone, in by far the plurality of instances, is more interesting than the surface of an ordinary hill; more fantastic in form, and incomparably richer in colour – the last quality being most noble in stones of good birth (that is to say, fallen from the crystalline mountain ranges).
J. Ruskin, *Modern Painters*, Vol. 5, Chapter 18 (1860)

The scale invariance of geological phenomena is one of the first concepts taught to a student of geology. It is pointed out that an object that defines the scale, i.e. a coin, a rock hammer, a person, must be included whenever a photograph of a geological feature is taken. Without the scale it is often impossible to determine whether the photograph covers 10 cm or 10 km. For example, self-similar folds occur over this range of scales. Another example would be an aerial photograph of a rocky coastline. Without an object with a characteristic dimension, such as a tree or house, the elevation of the photograph cannot be determined. It was in this context that Mandelbrot (1967) introduced the concept of fractals. The length of a rocky coastline is obtained using a measuring rod with a specified length. Because of scale invariance, the length of the coastline increases as the length of the measuring rod decreases according to a power law; the power determines the fractal dimension of the coastline. It is not possible to obtain a specific value for the length of a coastline, owing to all the small indentations down to a scale of millimeters or less.

Many geological phenomena are scale invariant. Examples include frequency–size distributions of rock fragments, faults, earthquakes, volcanic eruptions, mineral deposits, and oil fields. A fractal distribution requires that the number of objects larger than a specified size has a power-law dependence on the size. The empirical applicability of power-law statistics to geological phenomena was recognized long before the concept of fractals was conceived. A striking example is the Gutenberg–Richter relation for the frequency–magnitude statistics of earthquakes (Gutenberg and Richter, 1954). The proportionality factor in the relationship between the number of earthquakes and earthquake magnitude is known as the b-value. It has been recognized for nearly 50 years that, almost universally, $b = 0.9$. It is now accepted that the Gutenberg–Richter relationship is equivalent to a fractal relationship between the number of earthquakes and the characteristic size of the rupture; the value of the fractal dimension D is simply twice the b-value; typically $D = 1.8$ for distributed seismicity.

Power-law distributions are certainly not the only statistical distributions that have been applied to geological phenomena. Other examples include the normal distribution and the log-normal distribution. However, the power-law distribution is the only distribution that does not include a characteristic length scale. Thus the power-law distribution must be applicable to scale-invariant phenomena. If a specified number of events are statistically independent the central-limit theorem provides a basis for the applicability of the Gaussian distribution. Scale invariance provides a rational basis for the applicability of the power-law, fractal distribution. Fractal concepts can also be applied to continuous distributions; an example is topography. Mandelbrot (1982) has used fractal concepts to generate synthetic landscapes that look remarkably similar to actual landscapes. The fractal dimension is a measure of the roughness of the features. The Earth's topography is a composite of many competing influences. Topography is created by tectonic processes including faulting, folding, and flexure. It is modified and destroyed by erosion and sedimentation. There is considerable empirical evidence that erosion is scale invariant and fractal; a river network is a classic example of a fractal tree. Topography often appears to be complex and chaotic, yet there is order in the complexity. A standard approach to the analysis of a continuous function such as topography along a linear track is to determine the coefficients A_n

in a Fourier series as a function of the wavelength λ_n. If the amplitudes A_n have a power-law dependence on wavelength λ_n a fractal distribution may result. For topography and bathymetry it is found that, to a good approximation, the Fourier amplitudes are proportional to the wavelengths. This is also true for Brownian noise, which can be generated by the random walk process as follows. Take a step forward and flip a coin; if tails occurs take a step to the right and if heads occurs take a step to the left; repeat the process. The random walk can also be described as fractal Brownian noise. The divergence of the walk or signal increases in proportion to the square root of the number of steps. A spectral analysis of the random walk shows that the Fourier coefficients A_n are proportional to wavelength λ_n.

Many geophysical data sets have power-law spectra. These include surface gravity and magnetics as well as topography. Since power-law spectra are defined by two quantities, the amplitude and the slope, these quantities can be used to carry out textural analyses of data sets. The fractal structure can also be used as the basis for interpolation between tracks where data have been obtained.

The philosophy of fractals has been beautifully set forth by their inventor Benoit Mandelbrot (Mandelbrot, 1982). A comprehensive treatment of fractals from the point of view of applications has been given by Feder (1988).

Although fractal distributions would be useful simply as a means of quantifying scale-invariant distributions, it is now becoming evident that their applicability to geological problems has a more fundamental basis. Lorenz (1963) derived a set of nonlinear differential equations that approximate thermal convection in a fluid. This set of equations was the first to be shown to exhibit chaotic behavior. Infinitesimal variations in initial conditions led to first-order differences in the solutions obtained. This is the definition of chaos. The equations are completely deterministic; however, because of the exponential sensitivity to initial conditions, the evolution of a chaotic solution is not predictable. The evolution of the solution must be treated statistically and the applicable statistics are often fractal.

The most universal example of chaotic behavior is fluid turbulence. It has long been recognized that turbulent flows must be treated statistically and that the appropriate spectral statistics are fractal. Since the flows in the earth's core that generate the magnetic field are expected to be turbulent, it is not surprising that they are also chaotic. The random reversals of the earth's magnetic field are a characteristic of chaotic behavior. In fact, solutions of a parameterized set of dynamo equations proposed by Rikitake (1958) exhibited spontaneous reversals and were subsequently shown to be examples of deterministic chaos (Cook and Roberts, 1970).

Recursion relations can also exhibit chaotic behavior. The classic example is the logistic map studied by May (1976). This simple quadratic relation has an amazing wealth of behavior. As the single parameter in the equation is varied the period of the recursive solution doubles until the solution becomes fully chaotic. The Lyapunov exponent is the quantitative test of chaotic behavior; it is a measure of whether adjacent solutions converge or diverge. If the Lyapunov exponent is positive the adjacent solutions diverge and chaotic behavior results. The logistic map and similar recursion relations are applicable to population dynamics and other ecological problems. The logistic map also produces fractal sets.

Slider-block models have long been recognized as a simple analog for the behavior of a fault. The block is dragged along a surface with a spring and the friction between the surface and the block can result in the stick–slip behavior that is characteristic of faults. Huang and Turcotte (1990a) have shown that a pair of slider blocks exhibits chaotic behavior in a manner that is totally analogous to the chaotic behavior of the logistic map. This is evidence that the deformation of the crust associated with displacements on faults is chaotic and thus, is a statistical process. This is entirely consistent with the observation that earthquakes obey fractal statistics.

Nonlinearity is a necessary condition for chaotic behavior. It is also a necessary condition for scale invariance and fractal statistics. Historically continuum mechanics has been dominated by the applications of three linear partial differential equations. They have also provided the foundations of geophysics. Outside the regions in which they are created, gravitational fields, electric fields, and magnetic fields all satisfy the Laplace equation. The wave equation provides the basis for understanding the propagation of seismic waves. And the heat equation provides the basis for understanding how heat is transferred within the earth. All of these equations are linear and none generate solutions that are chaotic. Also, the solutions are not scale invariant unless scale-invariant boundary conditions are applied.

Critical phenomena are also associated with fractals and chaos. The renormalization group method is one approach to critical phenomena. In this approach a calculation is carried out on a simple model on a small scale and then is renormalized to larger and larger scales. One application is to the onset of permeability through a grid of elements that have a specified probability of being permeable. There is a sudden onset of flow at a critical value of the element permeability. At this onset the statistics of the connected elements is fractal.

The concept of self-organized criticality was introduced by Bak *et al.* (1988) and can be applied to the distribution of seismicity. The definition of self-organized criticality is that a natural system in a marginally stable state, when perturbed from this state, will evolve back to the state of marginal stability. In the critical state there is no natural length scale so that fractal statistics are applicable. A simple cellular-automata model illustrates self-organized criticality. The statistics of failure in this model are fractal and resemble the frequency–magnitude statistics of earthquakes.

CHAPTER TWO

Definition of a fractal set

Since its original introduction by Mandelbrot (1967), the concept of fractals has found wide applicability. It has brought together under one umbrella a broad range of pre-existing concepts from pure mathematics to the most empirical aspects of engineering. It is not clear that a single mathematical definition can encompass all these applications, but we will begin our quantitative discussion by defining a fractal set according to

$$N_n = \frac{C}{r_n^D} \tag{2.1}$$

where N_n is the number of objects (i.e. fragments) with a characteristic linear dimension r_n, C is a constant of proportionality, and D is the fractal dimension. The fractal dimension can be an integer, in which case it is equivalent to a Euclidean dimension. The Euclidean dimension of a point is zero, of a line segment is one, of a square is two, and of a cube is three. In general, the fractal dimension is not an integer but a fractional dimension; this is the origin of the term fractal.

We now illustrate why it is appropriate to refer to D as a fractal or fractional dimension by using a line segment of unit length. Several examples of fractals are illustrated in Figure 2.1. In Figure 2.1(a) the line segment of unit length is divided into two parts so that $r_1 = \frac{1}{2}$ and one part is retained so that $N_1 = 1$. The remaining segment is then divided into two parts so that $r_2 = \frac{1}{4}$ and again one is retained so that $N_2 = 1$. In order to determine D, (2.1) can be written as

$$D = \frac{\ln(N_{n+1}/N_n)}{\ln(r_n/r_{n+1})} \tag{2.2}$$

where ln is a logarithm to the base e. For the example considered, $\ln(N_2/N_1) = \ln 1 = 0$, $\ln(r_1/r_2) = \ln 2$, and $D = 0$, the Euclidean

6

dimension of a point. This construction can be extended to higher and higher orders, but at each order n, $N_n = 1$, so that $\ln(N_{n+1}/N_n) = \ln 1 = 0$. As the order approaches infinity the remaining line length approaches zero and it becomes a point. Thus the Euclidean dimension of a point, zero, is appropriate. The construction illustrated in Figure 2.1(b) is similar except that the line segment is divided into three parts so that $r_1 = \frac{1}{3}$ and one is retained so that $N_1 = 1$. At the next order of the construction $r_2 = \frac{1}{9}$ and again

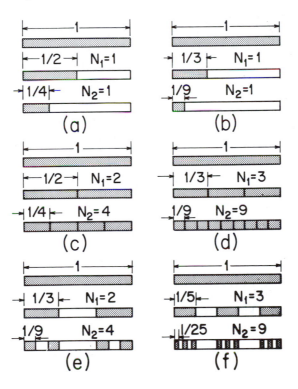

Figure 2.1. At order one a line segment of unit length is divided into an integer number of equal-sized smaller segments. A fraction of these segments is retained. The construction is repeated at higher orders. The first two orders are illustrated. (a) A line segment is divided into two parts and one is retained; $D = \ln 1/\ln 2 = 0$ (fractal dimension of a point). (b) A line segment is divided into three parts and one is retained; $D = \ln 1/\ln 3 = 0$ (fractal dimension of a point). (c) A line segment is divided into two parts and both are retained; $D = \ln 2/\ln 2 = 1$ (fractal dimension of a line). (d) A line segment is divided into three parts and all three are retained; $D = \ln 3/\ln 3 = 1$ (fractal dimension of a line). (e) A line segment is divided into three parts and two are retained; $D = \ln 2/\ln 3 = 0.6309$ (non-integer fractal dimension; this construction is also known as a Cantor set). (f) A line segment is divided into five parts and three are retained; $D = \ln 3/\ln 5 = 0.6826$ (non-integer fractal dimension).

$N_2 = 1$. Thus as the order is increased and $n \rightarrow \infty$ the construction again tends to a point and $D = 0$.

In Figure 2.1(c) the line segment of unit length is divided into two parts but both are retained so that $r_1 = \frac{1}{2}$ and $N_1 = 2$. The process is repeated so that $r_2 = \frac{1}{4}$ and $N_2 = 4$. From (2.2) we see that $D = \ln 2/\ln 2 = 1$. Similarly for Figure 2.1(d) we have $D = 1$; in both cases the fractal dimension is the Euclidean dimension of a line segment. This is appropriate since the remaining segment will be a line segment of unit length as the construction is repeated. However, not all constructions will give integer fractal dimensions; two examples are given in Figures 2.1(e) and 2.1(f). In Figure 2.1(e) the line segment of unit length is divided into three parts so that $r_1 = \frac{1}{3}$ and the two end segments are retained so that $N_1 = 2$. The process is repeated so that $r_2 = \frac{1}{9}$ and $N_2 = 4$. From (2.2) we find that $D = \ln 2/\ln 3 = 0.6309$. This is known as a Cantor set and has long been regarded by mathematicians as a pathological constuction. In Figure 2.1(f) the line segment is divided into five parts so that $r_1 = \frac{1}{5}$ and the two end segments and the center segment are retained: $N_1 = 3$. The process is repeated so that $r_2 = \frac{1}{25}$ and $N_2 = 9$. From (2.2) we find that $D = \ln 3/\ln 5 = 0.6826$. These two examples have fractal dimensions between the limiting cases of zero and one; thus they have fractional dimensions. Constructions can be devised to give any fractional dimension between zero and one using the method illustrated in Figure 2.1.

The iterative process illustrated in Figure 2.1 can be carried out as often as desired, making the remaining line lengths shorter and shorter. If n iterations are carried out then the line length at the nth iteration, r_n, is related to the length at the first iteration, r_1, by $r_n/r_0 = (r_1/r_0)^n$. Thus, as $n \rightarrow \infty$, $r_n \rightarrow 0$; in this limit the Cantor set illustrated in Figure 2.1(e) is known as a Cantor 'dust', an infinite set of clustered points. The repetitive iteration leading to a dust is known as 'curdling'. The scale invariance of the constructions given in Figure 2.1 is obvious. We refer to the nth iteration as order n. The line segment of length r_{n-1} at order n is scale invariant. As a particular example consider the Cantor set illustrated in Figure 2(e). The line segment of length one at order one has the middle third removed, the two remaining segments of length one-third at order two have the middle thirds removed, and so forth. Each line segment at order n is identical to the line segment at order one. This would also be true if the construction were extended to higher orders. Scale invariance is a necessary condition for the applicability of (2.1) since no natural length scale enters this power-law (fractal) relation.

The fractal concepts applied above to a line segment can also be applied to a square. A series of examples is given in Figure 2.2. In each case the square is divided into nine squares each with $r_1 = \frac{1}{3}$. Subsequently the remaining squares are divided into nine squares each with $r_2 = \frac{1}{9}$, and so forth. In Figure 2.2(a) only one square is retained, so that $N_1 = N_2 = N_n = 1$. From (2.2) $D = 0$, which is the Euclidian dimension of a point; this is appropriate since as $n \to \infty$ the remaining square will become a point. In Figure 2.2(b) two squares are retained at first order so that $r_1 = \frac{1}{3}$, $N_1 = 2$ and at second order $r_2 = \frac{1}{9}$, $N_2 = 4$. Thus from (2.2), $D = \ln 2/\ln 3 = 0.6309$, the same result that was obtained from Figure 2.1(e), as expected. Similarly, in Figure 2.2(c) three squares are retained so that $D = \ln 3/\ln 3 = 1$; in the limit $n \to \infty$ the remaining squares will become a line as in Figure 2.1(d). Thus the Euclidean dimension of a line is found. In Figure 2.2(d), only the center square is removed; thus at first order $r_1 = \frac{1}{3}$, $N_1 = 8$ and at second order $r_2 = \frac{1}{9}$, $N_2 = 64$. From (2.2) we have $D = \ln 8/\ln 3 = 1.8928$. This construction is known as a Sierpinski carpet. In Figure 2.2(e) all nine squares are retained; thus with $r_1 = \frac{1}{3}$, $N_1 = 9$ and with $r_2 = \frac{1}{9}$, $N_2 = 81$. From (2.2) we have $D = \ln 9/\ln 3 = 2$.

Figure 2.2. At order one the unit square is divided into nine equal-sized smaller squares with $r_1 = \frac{1}{3}$. At order two the remaining squares are divided into nine smaller equal-sized squares with $r_2 = \frac{1}{9}$. Five examples are given in which various numbers of squares, N, are retained. (a) $N_1 = 1$, $N_2 = 1$, $D = \ln 1/\ln 3 = 0$. (b) $N_1 = 2$, $N_2 = 4$, $D = \ln 2/\ln 3 = 0.6309$. (c) $N_1 = 3$, $N_2 = 9$, $D = \ln 3/\ln 3 = 1$. (d) $N_1 = 8$, $N_2 = 64$, $D = \ln 8/\ln 3 = 1.8928$ (also known as a Sierpinski carpet). (e) $N_1 = 9$, $N_2 = 81$, $D = \ln 9/\ln 3 = 2$.

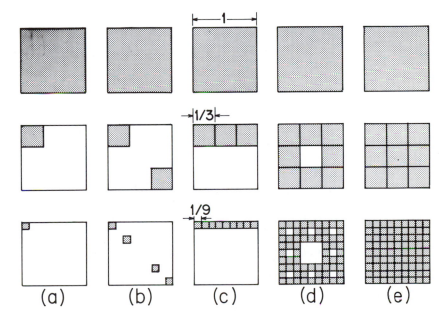

This is the Euclidean dimension of a square and is appropriate because when we retain all the blocks we continue to retain the unit square at all orders. Iterative constructions can be devised to yield any fractal dimensions between 0 and 2; again each construction is scale invariant.

The examples for one and two dimensions given in Figures 2.1 and 2.2 can be extended to three dimensions. Two examples are given in Figure 2.3. The Menger sponge is illustrated in Figure 2.3(a). A solid cube of unit dimensions has square passages with dimensions $r_1 = \frac{1}{3}$ cut through the centers of the six sides. The six cubes in the center of each side are removed as well as the central cube. Twenty cubes with dimensions $r_1 = \frac{1}{3}$ remain so that $N_1 = 20$. From (2.2) we find that $D = \ln 20/\ln 3 = 2.7268$. The Menger sponge can be used as a model for flow in a porous media with a fractal distribution of porosity. Another example of a fractal cube is given in Figure 2.3(b). Two solid cubes with dimensions $r_1 = \frac{1}{2}$ are removed from diagonally opposite corners; six solid cubes with dimensions $r_1 = \frac{1}{2}$ remain so that $N_1 = 6$. From (2.2) we find that $D = \ln 6/\ln 2 = 2.585$. We will use this configuration for a variety of applications in later chapters. Iterative constructions can be devised to yield any fractal dimension between 0 and 3; again each construction is scale invariant.

Figure 2.3. (a) At first order the unit cube is divided into 27 equal-sized smaller cubes with $r_1 = \frac{1}{3}$, 20 cubes are retained so that $N_1 = 20$. At second order $r_2 = \frac{1}{9}$ and 400 out of 729 cubes are retained so that $N_2 = 400$; $D = \ln 20/\ln 3 = 2.727$. This construction is known as the Menger sponge. (b) At first order the unit cube is divided into eight equal-sized smaller cubes with $r_1 = \frac{1}{2}$. Two diagonal opposite cubes are removed so that six cubes are retained and $N_1 = 6$. At second order $r_2 = \frac{1}{4}$ and 36 out of 64 cubes are retained so that $N_2 = 36$; $D = \ln 6/\ln 2 = 2.585$.

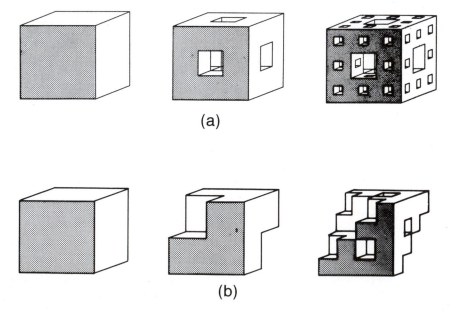

(a)

(b)

Figure 2.4. A fractal devil's staircase based on the third-order Cantor set illustrated in Figure 2.1(e). The horizontal step sizes are given by the Cantor set; the vertical step sizes are equal.

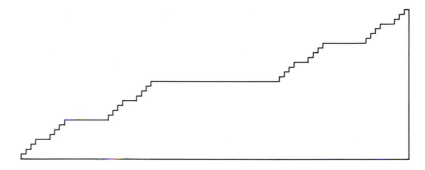

The examples given above illustrate how geometrical constructions can give non-integer, non-Euclidean dimensions. However, in each case the structure is not continuous. It is possible to use the Cantor set illustrated in Figure 2.1(e) to construct a continuous fractal. The result for a third-order Cantor set is given in Figure 2.4. Instead of removing the middle third of each line segment at each order, it is retained as a horizontal segment. The vertical segments are equal upward steps moving from left to right. This constuction is known as the devil's staircase and has the same fractal dimension as the Cantor set. It is similar to many wave-cut terraces on emergent coast lines.

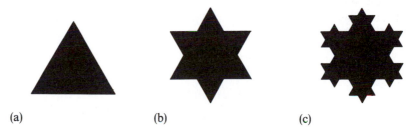

(a)   (b)   (c)

Figure 2.5. The triadic Koch island. (a) An equilaterial triangle with three sides of unit length. (b) Three triangles with sides of length $r_1 = \frac{1}{3}$ are placed in the center of each side. The perimeter is now made up of 12 sides and $N_1 = 12$. (c) Twelve triangles with sides of length $r_2 = \frac{1}{9}$ are placed in the center of each side. The perimeter is now made up of 48 sides and $N_2 = 48$; $D = \ln 4/\ln 3 = 1.26186$. The length of the perimeter in (a) is $P_0 = 3$, in (b) is $P_1 = 4$, and in (c) is $P_2 = \frac{16}{3} = 5.333$.

Another example of a continuous fractal construction is the triadic Koch island illustrated in Figure 2.5. This construction starts with an equilateral triangle with three sides of unit length, $N_0 = 3, r_0 = 1$. Equilateral triangles with sides of length $r_1 = \frac{1}{3}$ are placed in the center of each side. At order one there are 12 sides so that $N_1 = 12$. This construction is continued by placing equilateral triangles of length $r_2 = \frac{1}{9}$ in the center of each side. At order two there are 48 sides so that $N_2 = 48$. From (2.2) we have $D = \ln 4/\ln 3 = 1.26186$. The fractal dimension is between one (the Euclidian dimension of a line) and two (the Euclidean dimension of a surface). This construction can be continued to infinite order; the sides are scale invariant, a photograph of a side is identical at all scales. In order to quantify this we consider the length of the perimeter. The length of the perimeter P_n of a fractal island is given by

$$P_n = r_n N_n \tag{2.3}$$

where r_n is the side length at order n and N_n is the number of sides. Substitution of (2.1) gives

$$P_n = \frac{C}{r_n^{D-1}} \tag{2.4}$$

For the triadic Koch island illustrated in Figure 2.5 we have $P_0 = 3$, $P_1 = 4$, and $P_2 = \frac{16}{3} = 5.333$. Taking the logarithm of (2.4) and substituting these values we find that

$$D = 1 + \frac{\ln(P_{n+1}/P_n)}{\ln(r_n/r_{n+1})} = 1 + \frac{\ln 4/3}{\ln 3} = 1 + \frac{\ln 4 - \ln 3}{\ln 3} = \frac{\ln 4}{\ln 3} \tag{2.5}$$

This is the same result that was obtained above using (2.2), as expected. The perimeter of the triadic Koch island increases as n increases. As n approaches infinity the length of the perimeter also approaches infinity, as indicated by (2.4), since $D > 1$ (D is greater than unity). The perimeter of the triadic Koch island in the limit $n \to \infty$ is continuous but is not differentiable.

The triadic Koch island can be considered to be a model for measuring the length of a rocky coastline. However, there are several fundamental differences. The primary difference is that the perimeter of the Koch island is deterministic and the perimeter of a coastline is statistical. The perimeter of the Koch island is identically scale invariant at all scales. The perimeter of a rocky coastline will be statistically different at different scales but the differences do not allow the scale to be determined. Thus a rocky coastline is a statistical fractal. A second difference between the triadic Koch island and a rocky coastline is the range of scales over which scale invariance (fractal behavior) extends. Although a Koch island has the maximum scale of the origin triangle, the construction can be extended over an infinite range of scales. A rocky coastline has both a maximum scale and a minimum scale. The maximum scale would typically be 10^3 to 10^4 km, the size of the continent or island considered. The minimum scale would be the scale of the grain size of the rocks, typically 1 mm. Thus the scale invariance of a rocky coastline could extend over nine orders of magnitude. The existence of both upper and lower bounds is a characteristic of all naturally occurring fractal systems. Mandelbrot (1967) introduced the concept of fractals by using (2.4) to determine the fractal dimension of the west coast of Great Britain. The length of the coastline P_n was determined for a range of measuring rod lengths r_n. Mandelbrot (1967) used measurements of the length of the coastline obtained previously by Richardson (1961). Taking a map of a coastline, the length is obtained by using dividers of different lengths r_n. Using the scale of the map, the length of the coastline is plotted against the divider length on log–log paper. If the data points define a straight line the result is a statistical fractal. The result for the west coast of Great Britain is given in Figure 2.6. As shown, the data correlates well with (2.4), taking $D = 1.25$. This is evidence that the coastline is a fractal and is statistically scale invariant over this range of scales.

The technique for obtaining the fractal dimension of a coastline is easily extended to any topography. Contour lines on a topographic map are entirely equivalent to coastlines; the lengths along specified contours P_n are obtained using dividers of different lengths r_n. The fractal relation (2.4) is generally a good approximation and fractal dimensions can be obtained. As illustrated in Figure 2.7 the fractal

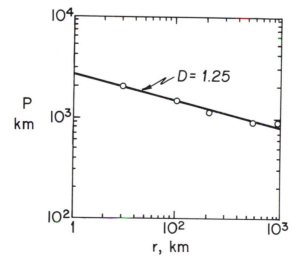

Figure 2.6. Length P of the west coast of Great Britain as a function of the length r of the measuring rod; data from Mandelbrot (1967). The data is correlated with (2.4) using $D = 1.25$.

dimensions of topography using the ruler method are generally in the range $D = 1.20 \pm 0.05$ independent of the tectonic setting and age. Topography is primarily a result of erosional processes; however, in young terrains topography is being created by active tectonic processes. It is not surprising that many of these processes are scale invariant and generate fractal topography. An interesting question, however, is whether erosional processes and tectonic processes each generate topographies with about the same fractal dimension.

It should be emphasized that not all topography is fractal (Goodchild, 1980). Young volcanic edifices are one example. Until modified by erosion, both shield and strata volcanoes are generally conical in shape and do not yield well defined fractal dimensions. Alluvial fans are another example of a non-fractal geomorphic feature. The morphology of alluvial fans can be modeled using the heat equation (Culling, 1960). Because the heat equation is linear it

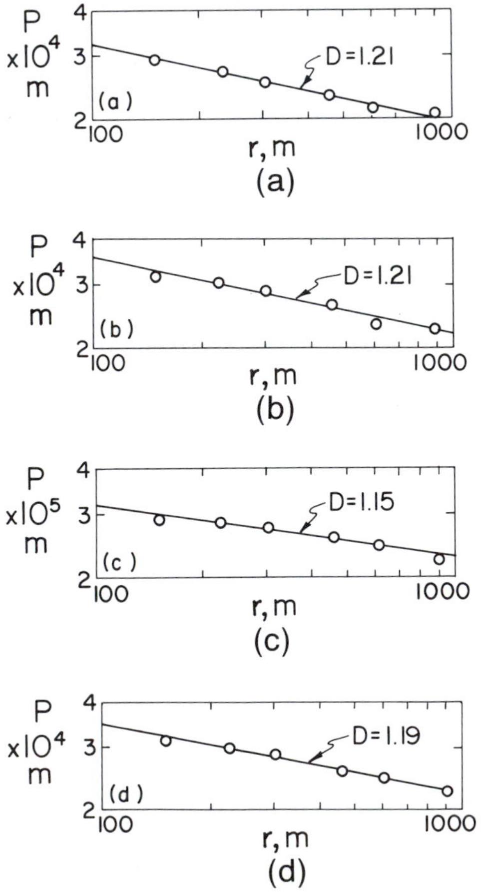

Figure 2.7. The lengths *P* of specified topographic contours in several mountain belts are given as functions of the length *r* of the measuring rod. (a) 3000 ft contour of the Cobblestone Mountain quadrangle, Transverse Ranges, California ($D = 1.21$); (b) 5400 ft contour of the Tatooh Buttes quadrangle, Cascade Mountains, Washington ($D = 1.21$); (c) 10 000 ft contour of the Byers Peak quadrangle, Rocky Mountains, Colorado ($D = 1.15$); (d) 1000 ft contour of the Silver Bay quadrangle, Adirondack Mountains, New York ($D = 1.19$).

contains a characteristic length (or time) and cannot give solutions that are scale invariant (fractal). The heat equation can also be used to model the elevation of mid-ocean ridges. The morphology of ocean trenches can be modeled by considering the bending of the elastic lithosphere. Again, the equation governing flexure is linear, introducing a characteristic length, and solutions are not scale invariant (fractal). However, despite these exceptions, most of the earth's topography and bathymetry is best modeled using fractal statistics and is therefore scale invariant.

Although the ruler method was the first used to obtain fractal dimensions it is not the most generally applicable method. The box-counting method has a much wider range of applicability than the ruler method (Pfeiffer and Obert, 1989). For example, it can be applied to a distribution of points as easily as it can be applied to a continuous curve. We now use the box-counting method to determine the fractal dimension of a rocky coastline. As a specific example we consider the coastline in the Dear Island, Maine quadrangle illustrated in Figure 2.8(a). The coastline is overlaid with a grid of square boxes; grids of different size boxes are used. The number of boxes N_n of size r_n required to cover the coastline is plotted on log–log paper as a function of r_n. If a straight line correlation is obtained then (2.2) is used to obtain the applicable fractal dimension. The box-counting method for the coastline given in Figure 2.8(a) is illustrated in Figures 2.8(b) and 2.8(c). The shaded areas are the boxes required to cover the coastline. In Figure 2.8(b) we require 98 boxes with $r = 1\,\mathrm{km}$ to cover the coastline; in Figure 2.8(c) we require 270 boxes with $r = 0.5\,\mathrm{km}$ to cover the coastline. The results for a range of box sizes are given in Figure 2.9. The correlation with (2.2) yields $D = 1.4$. This is somewhat higher than the values given above for other examples. But this is due to the extreme roughness of the coastline used in this example.

The statistical number–size distribution for a large number of objects can also be fractal. A specific example is rock fragments. In order that the distribution be fractal it is necessary that the number of objects N with a characteristic linear dimension greater than r should satisfy the relation

$$N = \frac{C}{r^D} \tag{2.6}$$

Figure 2.8. (a) Illustration
of a rocky coastline: the
Dear Island, Maine
quadrangle. (b) The shaded
area contains the square
boxes with $r = 1$ km
required to cover the
coastline; $N = 98$. (c) The
shaded area contains the
square boxes with
$r = 0.5$ km required to cover
the coastline; $N = 270$.

(a)

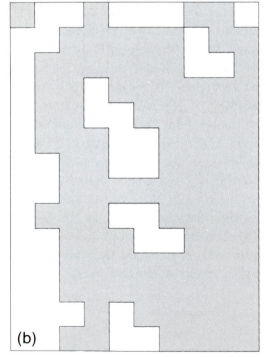

(b)

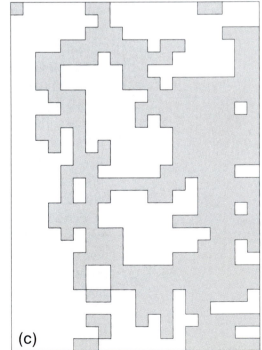

(c)

Figure 2.9. The number N of square boxes required to cover the coastline in Figure 2.8(a) as a function of the box size r. The correlation with (2.1) yields $D = 1.4$.

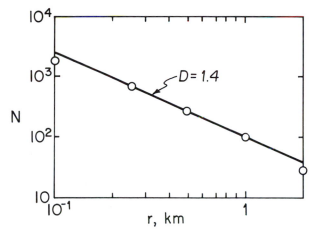

where D is again the fractal dimension. It is often appropriate to use this cumulative relation rather than the set relation (2.1). Another example where (2.6) is applicable is the frequency–size distribution of earthquakes.

As a statistical representation of a natural phenomenon (2.6) will be only approximately applicable, with both upper and lower bounds to the range of applicability. A specific example of the applicability of (2.6) is the Korcak empirical relation for the number of islands on the earth with an area greater than a specified value. Taking the characteristic length to be the square root of the area of the island, Mandelbrot (1975) showed that (2.6) is a good approximation with $D = 1.30$.

As we discussed, the term fractal dimension stands for fractional dimension. The meaning of this is clear in Figures 2.1–2.3; however, the meaning may be less clear in statistical power-law distributions. Some power-law distributions fall within the limits associated with fractional dimensions, i.e. $0 < D < 3$, but others do not. The question that must be addressed is whether all power-law distributions that satisfy (2.1) or (2.6) are fractal. In this book we make the assumption that they are fractal. Such distributions are clearly scale invariant even if not directly associated with a fractal dimension. This choice eliminates an ambiguity that can lead to considerable confusion when addressing measured data sets.

Problems

Problem 2.1. Consider the constuction illustrated in Figure 2.1(e). (a) Illustrate the construction at third order. (b) Determine N_3, N_4, r_3, and r_4.

Problem 2.2. Consider the construction illustrated in Figure 2.1(f). (a) Illustrate the construction at third order. (b) Determine N_3, N_4, r_3, and r_4.

Problem 2.3. A unit line segment is divided into five equal parts and two are retained. The construction is repeated. (a) Illustrate this construction to third order, i.e. consider $n = 1, 2, 3$. (b) Determine $N_1, N_2, N_3, r_1, r_2, r_3$. (c) Determine the fractal dimension.

Problem 2.4. A unit line segment is divided into seven equal parts and three are retained. The construction is repeated. (a) Illustrate this construction to second order, i.e. consider $n = 1, 2$. (b) Determine $N_1, N_2, N_3, r_1, r_2, r_3$. (c) Determine the fractal dimension.

Problem 2.5. A unit line segment is divided into seven equal parts and four are retained. The construction is repeated. (a) Illustrate this construction to second order, i.e. consider $n = 1, 2$. (b) Determine $N_1, N_2, N_3, r_1, r_2, r_3$. (c) Determine the fractal dimension.

Problem 2.6. Consider the construction of the Sierpinski carpet illustrated in Figure 2.2(d) at third order. (a) Illustrate the construction at third order. (b) Determine N_3, N_4, r_3, and r_4.

Problem 2.7. A unit square is divided into four smaller squares of equal size. Two diagonally opposite squares are retained and the construction is repeated. (a) Illustrate the construction to third order, i.e. consider $n = 1, 2, 3$. (b) Determine $N_1, N_2, N_3, r_1, r_2, r_3$. (c) Determine the fractal dimension.

Problem 2.8. A unit square is divided into 16 smaller squares of equal size. The four central squares are removed and the construction is repeated. (a) Illustrate this construction to second order, i.e. consider $n = 1, 2$. (b) Determine $N_1, N_2, N_3, r_1, r_2, r_3$. (c) Determine the fractal dimension.

Problem 2.9. A unit square is divided into 25 smaller squares of equal size. All squares are retained except the central one and the construction is repeated. (a) Illustrate this construction to second order, i.e. consider $n = 1, 2$. (b) Determine $N_1, N_2, N_3, r_1, r_2, r_3$. (c) Determine the fractal dimension.

Problem 2.10. A unit square is divided into 25 smaller squares of equal size. All the squares on the boundary and the central square are retained and the construction is repeated. (a) Illustrate this construction to second order, i.e. consider $n = 1, 2$. (b) Determine $N_1, N_2, N_3, r_1, r_2, r_3$. (c) Determine the fractal dimension.

Problem 2.11. A unit cube is divided into 27 smaller cubes of equal volume. All the cubes are retained except for the central one. What is the fractal dimension?

Problem 2.12. Construct a second-order devil's staircase based on the fractal construction given in Figure 2.1(f).

Problem 2.13. Consider a variation on the Koch island illustrated in Figure 2.5. At zero order again consider an equilateral triangle with three sides of unit length. At first order this triangle is enlarged so that it is an equilateral triangle with sides of length three. Equilateral triangles with sides of unit length are placed in the center of each side. (a) Illustrate this construction at second order. (b) Determine the areas to second order, i.e. obtain A_0, A_1, A_2. (c) Do the areas given in (b) satisfy the fractal condition (2.1)? If the answer is yes, what is the fractal dimension?

Problem 2.14. Assume that the open squares in the Sierpinski carpet illustrated in Figure 2.2(d) represent lakes. (a) Determine the numbers of lakes to the third order, i.e. obtain N_1, N_2, N_3 corresponding to r_1, r_2, r_3. (b) Do the numbers of lakes given in (a) satisfy the fractal condition (2.1)? If the answer is yes, what is the fractal dimension?

CHAPTER THREE

Fragmentation

As our first example of observed fractal distributions we consider fragmentation. Fragmentation plays an important role in a variety of geological phenomena. The earth's crust is fragmented by tectonic processes involving faults, fractures, and joint sets. Rocks are further fragmented by weathering processes. Rocks are also fragmented by explosive processes, both natural and man made. Volcanic eruptions are an example of a natural explosive process. Impacts produce fragmented ejecta. Although fragmentation is of considerable economic importance and many experimental, numerical, and theoretical studies have been carried out, relatively little progress has been made in developing comprehensive theories of fragmentation. A primary reason is that fragmentation involves the initiation and propagation of fractures. Fracture propagation is a highly nonlinear process requiring complex models for even the simplest configuration. Fragmentation involves the interaction between fractures over a wide range of scales. Fragmentation phenomena have been discussed by Grady and Kipp (1987) and Clark (1987). If fragments are produced with a wide range of sizes and if natural scales are not associated with either the fragmented material or the fragmentation process, fractal distributions of number versus size would seem to be expected. Some fractal aspects of fragmentation have been considered by Turcotte (1986a).

A variety of statistical descriptions has been used to represent the frequency–size distribution of fragments. An extensively used empirical description is the power-law relation

$$N(>m) = Cm^{-b} \tag{3.1}$$

where $N(>m)$ is the number of fragments with mass greater than *m*. The constants C and b are chosen to fit observed distributions; we now show that the empirical constant b is equivalent to the fractal dimension D. Since fragments can occur in a variety of shapes it is appropriate to define a linear dimension r as the cube root of volume: $r = V^{1/3}$. Assuming constant density it follows that $m \sim r^3$; a comparison of (3.1) with the fractal distribution defined in (2.6) gives

$$D = 3b \qquad (3.2)$$

Thus the power-law distribution (3.1) is equivalent to the fractal distribution (2.6).

An alternative empirical correlation for the size–frequency distribution is the Weibull distribution given by

$$\frac{M(<r)}{M_0} = 1 - \exp\left[-\left(\frac{r}{r_0}\right)^v\right] \qquad (3.3)$$

where $M(<r)$ is the cumulative mass of fragments with size less than r, M_0 is the total mass of fragments, and r_0 is related to their mean size. The power v is an arbitrary constant but is generally taken to be an integer. If $v = 2$ then (3.3) is a quadratic Weibull distribution; v is often taken to be a larger integer. Measurements of size distributions are often given in terms of mass. The mass of fragments smaller than a specified dimension is obtained directly from a sieve or screen analysis; the mass of fragments passing through a sieve with a specified aperture is obtained directly. For the total mass of fragments we can write

$$M(>r) + M(<r) = M_0 \qquad (3.4)$$

where $M(>r)$ is the cumulative mass of fragments with size greater than r. Combining (3.3) and (3.4) gives

$$\frac{M(>r)}{M_0} = \exp\left[-\left(\frac{r}{r_0}\right)^v\right] \qquad (3.5)$$

This is the Rosin and Rammler (1933) distribution law which is used extensively in geological applications. The Rosin and Rammler law (3.5) is entirely equivalent to the Weibull relation (3.3).

If $(r/r_0)^v$ is small then the exponential can be expanded in a Taylor series to give

$$\exp\left[-\left(\frac{r}{r_0}\right)^v\right] = 1 - \left(\frac{r}{r_0}\right)^v + \cdots \qquad (3.6)$$

where higher powers of $(r/r_0)^v$ have been neglected. Substitution of

(3.6) into the Weibull distribution (3.3) gives

$$\frac{M(<r)}{M_0} = \left(\frac{r}{r_0}\right)^{v} \tag{3.7}$$

Thus, for the small fragments, the Weibull distribution reduces to a power-law relation.

The power-law mass relation (3.7) can be related to the fractal number relation (2.6) by taking incremental values. Taking the derivative of (3.7) gives

$$dM \sim r^{v-1}\,dr \tag{3.8}$$

Taking the derivative of (2.6) gives

$$dN \sim r^{-D-1}\,dr \tag{3.9}$$

However, the incremental number is related to the incremental mass by

$$dN \sim r^{-3}\,dM \tag{3.10}$$

Substitution of (3.8) and (3.9) into (3.10) gives

$$r^{-D-1} \sim r^{-3}r^{v-1} \tag{3.11}$$

or

$$D = 3 - v \tag{3.12}$$

When data is obtained by sieve analyses, (3.12) is used to convert mass distributions to number distributions in order to specify a fractal dimension.

It is often convenient to specify the mass distribution with a distribution function. Taking the derivative of the Weibull relation (3.3) we obtain the distribution function

$$f_{\mathrm{w}}(r) = \frac{vr^{v-1}}{r_0^{v}}\exp\left[-\left(\frac{r}{r_0}\right)^{v}\right] \tag{3.13}$$

where $f(r)\,dr$ is the mass fraction of fragments with linear sizes between r and $r + dr$. The integral of $f(r)$ from $r = 0$ to ∞ must be unity since it includes all fragments. An alternative distribution function that is often used in geological problems is the log-normal distribution. A variable r has a log-normal distribution if the logarithms of its values are normally distributed. The log-normal distribution for fragmentation can be written

$$f_{\mathrm{ln}}(r) = \frac{1}{(2\pi)^{1/2}\sigma r}\exp\left[-\frac{1}{2}\left(\frac{\ln r - \mu}{\sigma}\right)^{2}\right] \tag{3.14}$$

where σ and μ are free parameters that can be related to the mean and the variance of the distribution. The mean radius $\bar{r}$ for fragments with a log-normal distribution is given by

$$\bar{r} = \int_0^\infty r f_{\ln}(r)\,dr = \exp(\mu + \tfrac{1}{2}\sigma^2) \qquad (3.15)$$

The variance V for the log-norml distribution is given by

$$V = \int_0^\infty (r - \bar{r})^2 f_{\ln}(r)\,dr = \exp(2\mu + \sigma^2)(\exp\sigma^2 - 1) \qquad (3.16)$$

The log-normal distribution is not scale invariant, but the Weibull distribution is scale invariant (i.e. it is a power-law distribution) for small r.

Many experimental studies of the frequency–size distributions of fragments have been carried out. Several examples of power-law fragmentation are given in Figure 3.1. A classic study of the frequency–size distribution for broken coal was carried out by Bennett (1936). The frequency–size distribution for the chimney rubble above the PILEDRIVER nuclear explosion in Nevada has been given by Schoutens (1979). This was a 61 kt event at a depth of 457 m in granite. The frequency–size distribution for fragments resulting from the high-velocity impact of a projectile on basalt has

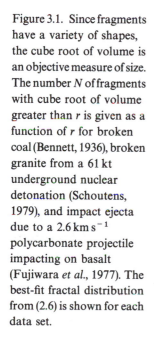

Figure 3.1. Since fragments have a variety of shapes, the cube root of volume is an objective measure of size. The number N of fragments with cube root of volume greater than r is given as a function of r for broken coal (Bennett, 1936), broken granite from a 61 kt underground nuclear detonation (Schoutens, 1979), and impact ejecta due to a 2.6 km s^{-1} polycarbonate projectile impacting on basalt (Fujiwara *et al.*, 1977). The best-fit fractal distribution from (2.6) is shown for each data set.

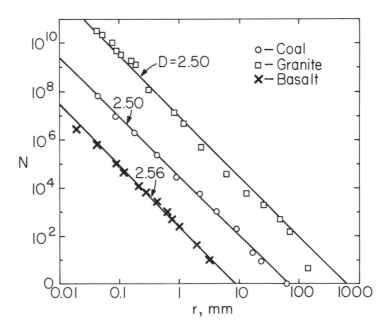

Table 3.1. *Fractal dimensions for a variety of fragmented objects*

Object	Reference	Fractal dimension D
Artificially crushed quartz	Hartmann (1969)	1.89
Disaggregated gneiss	Hartmann (1969)	2.13
Disaggregated granite	Hartmann (1969)	2.22
FLAT TOP I (chemical explosion, 0.2 kt)	Schoutens (1979)	2.42
PILEDRIVER (nuclear explosion, 62 kt)	Schoutens (1979)	2.50
Broken coal	Bennett (1936)	2.50
Projectile fragmentation of quartzite	Curran *et al.* (1977)	2.55
Projectile fragmentation of basalt	Fujiwara *et al.* (1977)	2.56
Fault gouge	Sammis and Biegel (1989)	2.60
Sandy clays	Hartmann (1969)	2.61
Terrace sands and gravels	Hartmann (1969)	2.82
Glacial till	Hartmann (1969)	2.88
Ash and pumice	Hartmann (1969)	3.54

been given by Fujiwara *et al.* (1977). In each of the three examples a good correlation with the fractal relation (2.6) is obtained over two to four orders of magnitude. In each example the fractal dimension for the distribution is near $D = 2.5$.

Further examples of power-law distributions for fragments are given in Table 3.1. It will be seen that a great variety of fragmentation processes can be interpreted in terms of a fractal dimension. Many examples involve fragmentation due to the weathering of rock (Hartmann, 1969). It has also been shown that fault gouge has a fractal frequency–size distribution (Sammis *et al.*, 1986; Sammis and Biegel, 1989). It is seen that the values of the fractal dimension vary considerably, but most lie in the range $2 < D < 3$. This range of fractal dimensions can be related to the total volume of fragments and to their surface area.

The total volume (mass) of fragments is given by

$$V = \int_{r_{\min}}^{r_{\max}} r^3 \, \mathrm{d}N \qquad (3.17)$$

since r has been defined to be the cube root of the volume. In all cases it is expected that there will be upper and lower limits to the validity of the fractal (power-law) relation for fragmentation. The upper limit r_{max} is generally controlled by the size of the object or region that is being fragmented. The lower limit r_{min} is likely to be controlled by the scale of the heterogeneities responsible for fragmentation, for example the grain size. For a power-law (fractal) distribution of sizes, substitution of (3.9) into (3.17) and integration gives

$$V \sim \frac{1}{3-D}(r_{max}^{3-D} - r_{min}^{3-D}) \tag{3.18}$$

If $0 < D < 3$ it is necessary to specify r_{max} but not r_{min} in order to obtain a finite volume (mass) of fragments. The volume (mass) of fragments is predominantly in the largest fragments. This is the case for most observed distributions of fragments (see Table 3.1). If $D > 3$ it is necessary to specify r_{min} but not r_{max}. The volume (mass) of the small fragments dominates.

The total surface area A of the fragments is given by

$$A = C \int_{r_{min}}^{r_{max}} r^2 \, dN \tag{3.19}$$

where C is a geometrical factor depending upon the average shape of the fragments. For a power-law distribution, substitution of (3.9) into (3.19) and integration gives

$$A \sim \frac{C}{D-2}\left(\frac{1}{r_{min}^{D-2}} - \frac{1}{r_{max}^{D-2}}\right) \tag{3.20}$$

If $0 < D < 2$ it is necessary to specify r_{max} but not r_{min} in order to obtain a finite total surface area for the fragments. But if $D > 2$ it is necessary to specify r_{min} in order to constrain the total surface area to a finite value. Thus for most observed distributions of fragments (see Table 3.1) the surface area of the smallest fragments dominates.

A simple model illustrates how fragmentation can result in a fractal distribution. This model is illustrated in Figure 3.2; it is based on the concept of renormalization, which will be considered in greater detail in Chapter 15. A cube with a linear dimension h is referred to as a zero-order cell; there are N_0 of these cells. Each zero-order cell may be divided into eight equal-sized, zero-order cubic elements with dimensions $h/2$. The volume V_1 of each of these elements is given by

$$V_1 = \tfrac{1}{8}V_0 \tag{3.21}$$

where V_0 is the volume of the zero-order cells. The probability that a zero-order cell will fragment to produce eight zero-order elements is taken to be f. The number of zero-order elements produced by fragmentation is

$$N_1 = 8fN_0 \tag{3.22}$$

After fragmentation the number of zero-order cells that have not been fragmented, N_{0a}, is given by

$$N_{0a} = (1 - f)N_0 \tag{3.23}$$

Each of the zero-order elements is now taken to be a first-order cell. Each first-order cell may be fragmented into eight equal-sized, first-order cubic elements with dimensions $h/4$. The fragmentation process is repeated for these smaller cubes. The problem is renormalized and the cubes with dimension $h/2$ are treated in exactly the same way that the cubes with linear dimension h were treated above. Each of the fragmented cubic elements with linear dimension $h/2$ is taken to be a first-order cell; each of these cells is divided into eight first-order cubic elements with linear dimensions $h/4$ as illustrated in Figure 3.2. The volume of each first-order element is

$$V_2 = \frac{1}{8}V_1 = \frac{1}{8^2}V_0 \tag{3.24}$$

Figure 3.2. Idealized model for fractal fragmentation. A zero-order cubic cell with dimensions h is divided into eight zero-order cubic elements each with dimensions $h/2$. The probability that a zero-order cell will be fragmented into eight zero-order elements is f. The fragments with dimensions $h/2$ become first-order cells; each of these have a probability f of being fragmented into first-order elements with dimensions $h/4$. The process is repeated to higher orders. The basic structure is fractal.

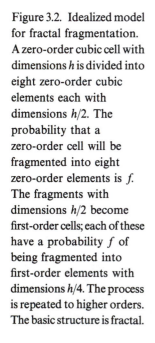
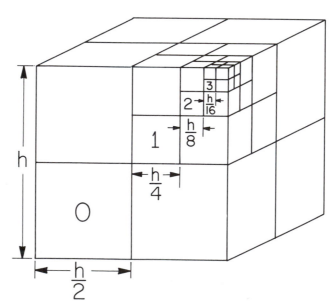

The probability that a first-order cell will fragment is again taken to be f in order to preserve scale invariance. The number of fragmented first-order elements is

$$N_2 = 8fN_1 = (8f)^2 N_0 \tag{3.25}$$

After fragmentation the number of first-order cells that have not been fragmented is

$$N_{1a} = 8f(1-f)N_0 \tag{3.26}$$

This process is repeated at successively higher orders. The volume of the nth-order cell V_n is given by

$$V_n = \frac{1}{8^n} V_0 \tag{3.27}$$

After fragmentation the number of nth-order cells is

$$N_{na} = (8f)^n(1-f)N_0 = (8f)^n N_{0a} \tag{3.28}$$

Taking the natural logarithm of both sides we can write (3.27) and (3.28) as

$$\ln\left(\frac{V_n}{V_0}\right) = -n \ln 8 \tag{3.29}$$

$$\ln\left(\frac{N_{na}}{N_{0a}}\right) = n \ln 8f \tag{3.30}$$

Elimination of n from (3.29) and (3.30) gives

$$\frac{N_{na}}{N_{0a}} = \left(\frac{V_n}{V_0}\right)^{-\ln 8f/\ln 8} \tag{3.31}$$

Comparison with (2.1) shows that this is a fractal distribution with

$$D = 3\frac{\ln 8f}{\ln 8} \tag{3.32}$$

Although this model is very idealized and non-unique, it illustrates the basic principles of how scale-invariant fragmentation leads to a fractal distribution. It also illustrates the principle of renormalization. The division into eight fragments is an arbitrary choice, however; other choices such as the division into two or 16 fragments will give the same result. This model is deterministic rather than statistical. Actual distributions of fragments are continuous rather than discrete but the deterministic model can be related to a 'bin' analysis of a statistical distribution. Also, this model relates the probability of fragmentation f to the fractal dimension D but does not place constraints on the value of the fractal dimension.

It is of interest to discuss this model in terms of the allowed range of D. The allowed range of f is $\frac{1}{8} < f < 1$ and the equivalent range of D is $0 < D < 3$. Thus the concept of fractional dimension introduced in Figure 2.3 appears to be appropriate for fragmentation. However, the data for artificially crushed quartz given in Table 3.1 falls outside this allowed range. Since such distributions are not precluded physically, we will consider this a fractal (scale-invariant) distribution even though it lies outside the geometrically allowed range. We accept the physical view rather than the mathematical view.

Fragmentation is a process with a wide range of applications. Thus many studies have been carried out to prescribe size distributions in terms of basic physics; however, fragmentation is a very complex problem. The results given above indicate that in many cases fragmentation is a scale-invariant process that leads to a fractal distribution. We now turn to a discrete model of fragmentation that does yield a specific fractal dimension. We will consider the fractal cube illustrated in Figure 2.3(b) and use it as a basis for a fragmentation model. This model is illustrated in Figure 3.3. Although the geometry and fractal dimension are the same, the concepts of the two models are quite different. The models given in Figure 2.3 are essentially for a porous (Swiss cheese) configuration. At each scale blocks are removed to create void space. In this chapter we consider fragmentation such that some blocks are retained at each scale but

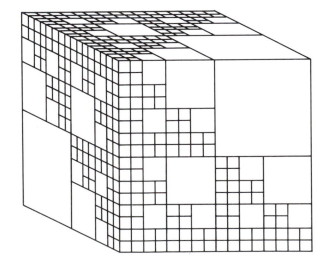

Figure 3.3. Illustration of a fractal model for fragmentation. Two diagonally opposite cubes are retained at each scale. With $r_1 = h/2$, $N_1 = 2$ and $r_2 = h/4$, $N_2 = 12$ we have $D = \ln 6/\ln 2 = 2.5850$.

others are fragmented. In the model given in Figure 3.3 two diagonally opposed blocks are retained at each scale. No two blocks of equal size are in direct contact with each other. This is the comminution model for fragmentation proposed by Sammis *et al.* (1986). It is based on the hypothesis that direct contact between two fragments of near equal size during the fragmentation process will result in the breakup of one of the blocks. It is unlikely that small fragments will break large fragments or that large fragments will break small fragments.

For the configuration illustrated in Figure 3.3 we have $N_1 = 2$ for $r_1 = h/2$, $N_2 = 12$ for $r_2 = h/4$, and $N_3 = 72$ for $r_3 = h/8$. From (2.2) we find that $D = \ln 6/\ln 2 = 2.5850$. This is the fractal distribution of a discrete set but we wish to compare it with statistical fractals obtained from the actual fragmentation observations. It is therefore of interest to consider also the cumulative statistics for the comminution model. The cumulative number of blocks larger than a specified size for the three highest orders are $N_{1c} = 2$ for $r_1 = h/2$, $N_{2c} = 14$ for $r_2 = h/4$, and $N_{3c} = 86$ for $r_3 = h/8$; N_{nc} is the cumulative number of the fragments equal to or larger than r_n. The cumulative statistics for the model illustrated in Figure 3.3 are given in Figure 3.4; excellent agreement with the fractal relation (2.6) is obtained taking $D = 2.60$. Thus the fractal dimensions for the discrete set and the cumulative statistics are nearly equal.

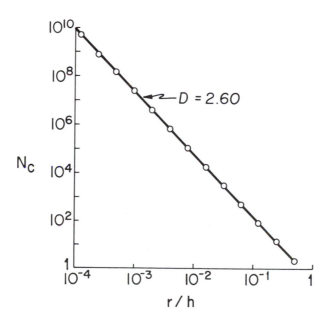

Figure 3.4. Cumulative statistics for the fragmentation model illustrated in Figure 3.3. Correlation with (2.6) gives $D = 2.60$.

The comminution model was originally developed for fault gouge. The relative displacement across a fault zone results in the fragmentation of the wall rock to form a zone of fragmented rock known as fault gouge. Sammis and Biegel (1989) have shown that fault gouge obtained from the Lopez fault zone, San Gabriel Mountains, California has a fractal dimension of $D = 2.60 \pm 0.11$ on scales from 0.5 μm to 10 mm. Synthetic fault gouge has been produced in the laboratory and has also been shown to have a fractal dimension $D = 2.60$ (Biegel *et al.*, 1989). In both cases the results are in excellent agreement with the comminution model, which gave $D = 2.60$. It is seen from Figure 3.1 and Table 3.1 that many observed distributions of fragments have fractal distributions near this value. This is evidence that the comminution model may be widely applicable to rock fragmentation.

The comminution model may also be applicable to tectonic zones in the earth's crust. The tectonic zone of active seismicity that results when two surface plates slide past one another often results in a zone of fragments similar to fault gouge, but on a larger scale. This model will be discussed further in the next chapter.

Most rock has a natural porosity. This porosity often provides the necessary permeability for fluid flow. There are generally two types of porosity, intergranular porosity and fracture porosity. Based on the discussion given above it would not be surprising if both types of porosity exhibited fractal behavior. Fractures are directly related to fragmentation, and granular rocks such as sandstone are composed of rock fragments with a variety of scales. Based on laboratory studies a number of authors have suggested that sandstones have a fractal distribution of porosity (Katz and Thompson, 1985; Krohn and Thompson, 1986; Daccord and Lenormand, 1987; Krohn, 1988a, b; Thompson *et al.*, 1987).

We previously introduced models with scale invariant porosity in Figure 2.3. The Menger sponge, Figure 2.3(a), can be taken as a simple model for a porous medium. However, this model is conceptually somewhat different to that considered in Chapter 2. The model is constructed from solid cubes of density ρ_0 and size r_0. We construct a first-order Menger sponge from these cubes; the size of the first-order cube is $r_1 = 3r_0$. The first-order sponge is made up of 20 solid zero-order cubes so that the first-order porosity is $\phi_1 = 7/27$ and the first-order density is $\rho_1 = 20\rho_0/27$. Continuing the construction to second order the size of the cube is $r_2 = 9r_0$ and there are 400 solid cubes of size r_0 with density ρ_0. Thus the porosity of the second-order Menger sponge is $\phi_2 = 329/729$ and its density is $\rho_2 = 400\rho_0/729$. The porosity of the nth-order Menger sponge is

$$\phi = 1 - \left(\frac{r_0}{r_n}\right)^{3 - \ln 20/\ln 3} \tag{3.33}$$

which is not a power-law (fractal) relation. The density of the nth-order Mengor sponge is

$$\frac{\rho_n}{\rho_0} = \left(\frac{r_0}{r_n}\right)^{3 - \ln 20/\ln 3} \tag{3.34}$$

This is a fractal relation and it is illustrated in Figure 3.5. For the Mengor sponge the fractal dimension is $D = \ln 20/\ln 3 = 2.727$. Generalizing (3.33), the porosity ϕ for a fractal medium can be related to its fractal dimension by

$$\phi = 1 - \left(\frac{r_0}{r}\right)^{3 - D} \tag{3.35}$$

where r is the linear dimension of the sample considered. Similarly, the density of the fractal medium scales with its size according to

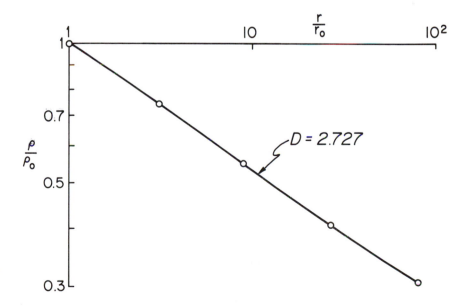

Figure 3.5. Density dependence ρ/ρ_0 of a Menger sponge as a function of the size of the sponge r/r_0. A fractal decrease in the density is found with $D = 2.727$ from (3.34).

$$\frac{\rho}{\rho_0} = \left(\frac{r_0}{r}\right)^{3-D} \tag{3.36}$$

The density of a fractal solid systematically decreases with the increasing size of the sample considered.

A number of studies of the densities of soil aggregates as a function of size have been carried out. These studies show a systematic decrease in density as the size of the aggregate increases. A sieve analysis is carried out on a soil and the mean density of each aggregate is found. The results for a sandy loam obtained by Chepil (1950) are given in Figure 3.6. Although there is scatter, the results agree reasonably well with the fractal soil porosity from (3.36) using as the fractal dimension $D = 2.869$.

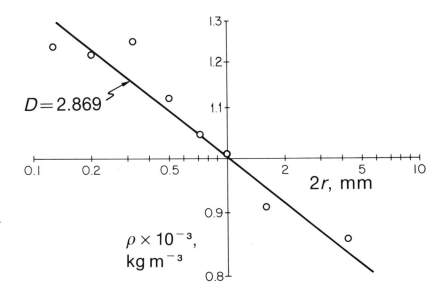

Figure 3.6. Density of soil aggregates as a function of their size (Chepil, 1950). The solid line is from (3.34) with $D = 2.869$.

Problems

Problem 3.1. The distribution function for the power-law mass distribution given by (3.7) is

$$f_p = v \frac{r^{v-1}}{r_0^v}$$

Assume that the maximum fragment size is r_0 and that $v > 1$. Determine the mean fragment radius $\bar{r}$ and the variance V about this mean.

Problem 3.2. Consider a bar of unit length that has a probability f_2 of being fragmented into two bars of equal length 1/2. The two smaller bars have the same probability of being fragmented into bars of length 1/4. Show that this process leads to a fractal distribution with

$$D = 3 \frac{\ln 2f^2}{\ln 2}$$

Show that this result is equivalent to (3.32).

Problem 3.3. Consider a cube with a linear dimension h that is divided into 64 cubic elements with a dimension of $h/4$. The probability of fragmentation is f_{64}. The smaller cubes have the same probability of being fragmented into cubes with dimensions of $h/16$. Show that this process leads to a fractal distribution with

$$D = 3 \frac{\ln(64f_{64})}{\ln 64}$$

Show that this result is equivalent to (3.32).

Problem 3.4. Consider the fragmentation model illustrated in Figure 3.7.

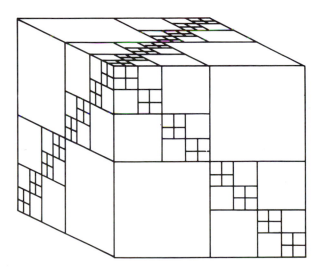

Figure 3.7. Illustration of a fractal model for fragmentation. Four cubic elements are retained at each scale.

Determine N_1 for $r_1 = h/2$, N_2 for $r_2 = h/4$, and N_3 for $r_3 = h/8$. What is the fractal dimension of the fragments?

Problem 3.5. A model for a porous medium is constructed from solid cubes of density ρ_0 and size r_0. At first order 26 of these cubes are used to construct a cube with $r_1 = 3r_0$, the central cube is missing. Determine the porosity ϕ_1 and the density ρ_1. Continue the construction and determine ϕ_2 and ρ_2. What is the fractal dimension?

Problem 3.6. A model for a porous medium is constructed from solid cubes of density ρ_0 and size r_0. At first order 21 of these cubes are used to construct a cube with $r_1 = 3r_0$; the cubes in the center of each face are missing (i.e. a Mengor sponge with a central cube). Determine porosity ϕ_1 and the density ρ_1. Continue the construction and determine ϕ_2 and ρ_2. What is the fractal dimension?

CHAPTER FOUR

Seismicity and tectonics

According to the hypothesis of plate tectonics, crustal deformation takes place at the boundaries between the major surface plates. In the idealized plate tectonic model plate boundaries are spreading centers (ocean ridges), subduction zones (ocean trenches), and transform faults (such as the San Andreas fault in California). Relative displacements at subduction zones and transform faults would occur on well defined faults. Displacements across these faults would be associated with great earthquakes such as the 1906 San Francisco earthquake. However, crustal deformation is more complex and is usually associated with relatively broad zones of deformation. Take the western United States as an example: although the San Andreas fault is the primary boundary between the Pacific and the North American plates, significant deformation takes place as far east as the Wasatch Front in Utah and the Rio Grande Graben in New Mexico. Active mountain building is occurring throughout the western United States. Distributed seismicity is associated with this mountain building. Even the displacements associated with the San Andreas fault system are distributed over many faults. South of San Francisco and north of Los Angeles the San Andreas fault has significant bends. Deformation associated with these bends is responsible for considerable mountain building and the 1956 Kern County earthquake, the 1971 San Fernando earthquake, and the 1989 Loma Prieta earthquake.

Although the crustal deformation in the western United States may appear to be complex, it does obey fractal statistics in a variety of ways. This is true of all zones of tectonic deformation. We will first consider the frequency–magnitude statistics of earthquakes.

Many regions of the world have dense seismic networks that can monitor earthquakes as small as magnitude two or less. The global seismic network is capable of monitoring earthquakes that occur anywhere in the world with a magnitude greater than about four.

Various statistical correlations have been used to relate the frequency of occurrence of earthquakes to their magnitude, but the most generally accepted is the log-linear relation (Gutenberg and Richter, 1954)

$$\log \dot{N} = -bm + \log \dot{a} \tag{4.1}$$

where b and $\dot{a}$ are constants, the logarithm is to the base 10, and $\dot{N}$ is the number of earthquakes per unit time with a magnitude greater than m occurring in a specified area. The Gutenberg–Richter law (4.1) is often written in terms of N, the number of earthquakes in a specified time interval (say 30 years), and the corresponding constant a.

The magnitude of an earthquake is a measure of its size based on the amplitude of the ground motions at seismographs at specified distances from the rupture. In this discussion we use the surface-wave magnitude based on the amplitude of surface waves at a period of 20 seconds. The Richter magnitude scale is a popular measure of the strength of earthquakes because of the logarithmic basis which allows essentially all earthquakes to be classified on a scale of 0–10. Alternative magnitude definitions include the local magnitude and the magnitude determined from body waves.

The frequency–magnitude relation (4.1) is found to be applicable over a wide range of earthquake sizes both globally and locally. The constant b or 'b-value' varies from region to region but is generally in the range $0.8 < b < 1.2$ (Evernden, 1970). The constant $\dot{a}$ is a measure of the regional level of seismicity.

The magnitude is an empirical measure of the size of an earthquake. It can be related to the total energy in the seismic waves generated by the earthquake, E_s, using the relation

$$\log E_s = 1.44m + 5.24 \tag{4.2}$$

where E_s is in Joules. The strain released during an earthquake is directly related to the moment M of the earthquake by the definition

$$M = \mu A \delta_e \tag{4.3}$$

where μ is the shear modulus of the rock in which the fault is embedded, A is the area of the fault break, and δ_e is the mean displacement across the fault during the earthquake. The moment of an earthquake can be related to its magnitude by

$$\log M = cm + d \qquad (4.4)$$

where c and d are contants. Kanamori and Anderson (1975) have established a theoretical basis for taking $c = 1.5$. Kanamori (1978) and Hanks and Kanamori (1979) have argued that (4.4) can be taken as a definition of magnitude with $c = 1.5$ and $d = 9.1$ (M in joules). This definition is consistent with the definitions of local magnitude and surface wave magnitude but not with the definition of body wave magnitude.

Kanamori and Anderson (1975) have also shown that it is a good approximation to relate the moment of an earthquake to the area A of the rupture by

$$M = \alpha A^{3/2} \qquad (4.5)$$

where α is a constant. Combining (4.1), (4.4), and (4.5) gives

$$\log \dot{N} = -\frac{3b}{2c} \log A + \log \dot{\beta} \qquad (4.6)$$

with

$$\log \dot{\beta} = \frac{bd}{c} + \log \dot{a} - \frac{b}{c} \log \alpha \qquad (4.7)$$

and (4.6) can be rewritten as

$$\dot{N} = \dot{\beta} A^{-3b/2c} \qquad (4.8)$$

In a specified region the number of earthquakes $\dot{N}$ per unit time with rupture areas greater than A has a power-law dependence on the area. A comparison with the definition of a fractal given in (2.6) with $A \sim r^2$ shows that the fractal dimension of distributed seismicity is

$$D = \frac{3b}{c} \qquad (4.9)$$

Taking the theoretical relation $c = 1.5$ we have

$$D = 2b \qquad (4.10)$$

Thus the fractal dimension of regional or world-wide seismic activity is simply twice the b-value. The empirical frequency–magnitude relation given in (4.1) is entirely equivalent to a fractal distribution (Aki, 1981).

The Gutenberg-Richter frequency-magnitude relation (4.1) has been found to be applicable under a great variety of circumtances. We will first consider its validity on a world-wide basis. Since 1920 the global seismic network has been able to monitor earthquakes with magnitudes greater than about seven. Purcaru and Berckhemer (1982) have given the moment of major earthquakes between 1920 and 1979. With improved instrument sensitivity and more stations, global monitoring was extended to magnitude five earthquakes in the 1980s. Dziewonski *et al.* (1989) and others referenced there have given moments for a broader range of earthquakes for the period 1983–87. Using (4.4) with $c = 1.5$ and $d = 9.1$ (M in joules) to convert moments to magnitudes, the frequency–magnitude statistics for earthquakes are given on a worldwide basis in Figure 4.1. The data correlate well with (4.1) taking $b = 1$ $(D = 2)$ and $\dot{a} = 10^8 \, \mathrm{yr}^{-1}$. Also given in Figure 4.1 is the equivalent characteristic length $A^{1/2}$ obtained from (4.5) taking $\alpha = 3.27 \times 10^6 \, \mathrm{Pa}$. The data given in Figure 4.1 can be used to estimate the frequency of occurrence of earthquakes of various magnitudes on a world-wide basis. For example, about 10 magnitude seven earthquakes are expected each year.

Several aspects of Figure 4.1 merit discussion. Clearly there must be an upper limit to the size of an earthquake. The San Andreas fault in California is nearly vertical and the maximum depth of

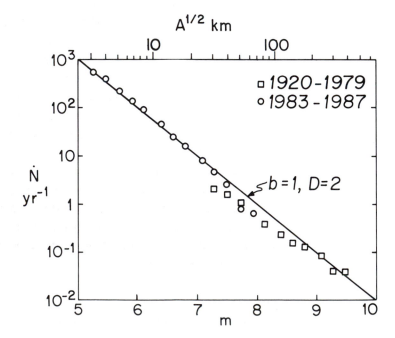

Figure 4.1. World-wide number of earthquakes per year, $\dot{N}$ with magnitudes greater than m as a function of m. The square root of the rupture area A is also given. Circles, 1983–1987 (Dziewonski *et al.*, 1989, and others); squares, 1920–1979 (Purcaru and Berckhemer, 1982). The solid line represents (4.1) with $b = 1.00$ and $\dot{a} = 10^8 \, \mathrm{yr}^{-1}$.

rupture is about 12 km. Plate boundary faults in subduction zones typically dip at about 20° and may have a vertical rupture length as great as 40 km in a single event. There must be a maximum horizontal rupture length of, say, 4000 km. This corresponds to $A^{1/2} = 400$ km which is consistent with the largest earthquake in Figure 4.1. Thus it is expected that there is an upper limit on the global fractal relation at around magnitude 9.5. Also, rupture geometries change from near equidimensional to elongated at about $A^{1/2} = 20$ km. This is the result of the depth limitation on rupture; large earthquakes have elongated rupture zones. There is no evidence for a change in the global b-value or fractal dimension at this transition.

It is of interest to consider the distribution of seismicity associated with the relative velocity v across a plate boundary. We consider a specified length of the fault zone which also has a specified depth of rupture. Thus the two plates are assumed to interact over an area A_p. The relative plate velocity v and interaction area A_p are related to the rupture area A and mean slip displacement δ_e in an individual earthquake by

$$vA_p = -\int A\delta_e \, d\dot{N} = \frac{-1}{\mu} \int M \, d\dot{N} \qquad (4.11)$$

where the integral is carried out over the entire distribution of seismicity and $d\dot{N}$ is the number of earthquakes per unit time with magnitudes between m and $m + dm$. The earthquake moment has been introduced from (4.3). We hypothesize that a fractal distribution of seismicity accommodates this relative velocity. From (4.1) and (4.4) we have

$$\dot{N} = \dot{a}10^{-bm} \qquad (4.12)$$

and

$$M = 10^d 10^{cm} \qquad (4.13)$$

Taking the derivative of (4.12) gives

$$d\dot{N} = -b\dot{a}(\ln 10)10^{-bm} \, dm \qquad (4.14)$$

Substitution of (4.13) and (4.14) into (4.11) gives

$$vA_p = \frac{b\dot{a}(\ln 10)10^d}{\mu} \int_{-\infty}^{m_{max}} 10^{(c-b)m} \, dm \qquad (4.15)$$

Since $c > b$ the integral diverges for large m so that the maximum-magnitude earthquake m_{max} must be specified. This is the well known observation that a large fraction of the total moment and energy associated with seismicity occurs in the largest events. Integration of (4.15) gives

$$vA_p = \frac{\dot{a}b}{\mu(c-b)} 10^{d(c-b)m_{max}} \tag{4.16}$$

A large value of regional strain vA_p implies either a high level of regional seismicity (large $\dot{a}$) or a large magnitude for the maximum-magnitude earthquake (large m_{max}).

This type of relation has been derived by several authors (Smith, 1976; Molnar, 1979; Anderson and Luco, 1983) and has been used to estimate regional strain (Anderson, 1986; Youngs and Coppersmith, 1985) and to compare levels of seismicity with known strain rates (Hyndman and Weichert, 1983; Singh *et al.*, 1983).

As a specific application we consider the regional seismicity in southern California. The frequency–magnitude distribution of seismicity in southern California as summarized by Main and Burton (1986) is given in Figure 4.2. Based on data from 1932 to 1972 the number of earthquakes per year $\dot{N}$ is given as a function of magnitude m (open squares). In the magnitude range $4.25 < m < 6.5$ the data is in excellent agreement with (4.1) taking $b = 0.89$ ($D = 1.78$) and $\dot{a} = 1.4 \times 10^5 \, \text{yr}^{-1}$. In terms of the linear dimension of the fault break, this magnitude range corresponds to a size range $1 \, \text{km} < A^{1/2} < 12 \, \text{km}$.

Also included in Figure 4.2 is the value of $\dot{N}$ associated with great earthquakes on the southern section of the San Andreas fault. Dates for 10 large earthquakes on this section of the fault have been obtained from radiocarbon dating of faults, folds, and liquefaction features within the marsh and stream deposits on Pallett Creek where it crosses the San Andreas fault 55 km northeast of Los Angeles (Sieh *et al.*, 1989). The mean repeat time is 132 years, giving $\dot{N} = 0.0076 \, \text{yr}^{-1}$. The most recent in the sequence of earthquakes occurred in 1857 and the observed offset across the fault associated with this earthquake was 12 m (Sieh and Jahns, 1984). Taking the inferred magnitude to be $m = 8.05$ the recurrence statistics for these large earthquakes are shown by the solid circle in Figure 4.2. An extrapolation of the fractal relation for regional seismicity appears to make a reasonable prediction of great earthquakes on this section of the San Andreas fault. Since this extrapolation is based on the 40 years of data between 1932 and 1972, a relatively small fraction of the mean interval of 132 years, it suggests that the value of $\dot{a}$ for this region may not have a strong dependence on time during the earthquake cycle. This conclusion has a number of important implications. If a great earthquake substantially relieved the regional stress then it would be expected that the regional seismicity would systematically increase as the stress increased before the next great earthquake. An alternative hypothesis is that an active tectonic zone is continuously in a critical state and that the fractal frequency–magnitude statistics are evidence for this critical behavior. In the critical state the background seismicity, small earthquakes not associated with aftershocks, have little time dependence. This hypothesis will be discussed in Chapter 16. Acceptance of this hypothesis allows the regional background seismicity to be used in assessing seismic hazards (Turcotte, 1989b). The regional frequency–magnitude statistics can be extrapolated to estimate recurrence times for larger magnitude earthquakes. Unfortunately, no information is provided on the largest earthquake to be expected.

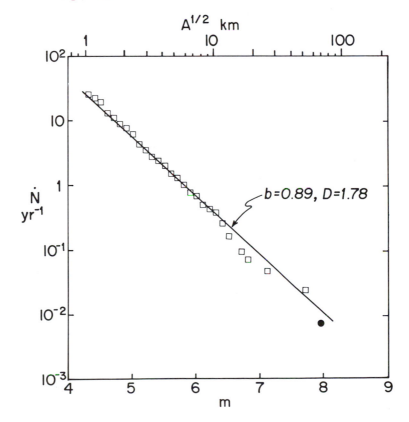

Figure 4.2. Number of earthquakes per year $\dot{N}$ with magnitudes greater than m as a function of m. Squares, Southern California 1932–1972 (Main and Burton, 1986); solid circle, expected rate of occurrence of great earthquakes in Southern California (Sieh et al., 1989). The line represents (4.1) with $b = 0.89$ and $\dot{a} = 1.4 \times 10^5 \, \text{yr}^{-1}$.

As an example of this approach let us consider seismicity in the eastern United States. Since the eastern United States is a plate interior the concept of rigid plates would preclude seismicity in the region. However, the plates act as stress guides. The forces that drive plate tectonics are applied at plate boundaries. The negative buoyancy force acting on the descending plate at a subduction zone acts as a 'trench pull'. Gravitational sliding off an ocean ridge acts as a 'ridge push'. Because the plates are essentially rigid these forces are transmitted through their interiors. However, the plates have zones of weakness that will deform under these forces and earthquakes result. Chinnery (1979) compiled frequency–magnitude statistics for three areas in the eastern United States and his results are given in Figure 4.3. Similar results for the Mississippi Valley region were obtained by Johnson and Nava (1985) based on instrumental and historical records. The Mississippi Valley region includes the site of the great New Madrid, Missouri earthquakes of 1811–12. This was a sequence of three earthquakes that were felt throughout the eastern United States and that disrupted the flow of the Mississippi River. Because they occurred prior to the invention of the seismograph and because no well defined fault break has been found it is difficult to

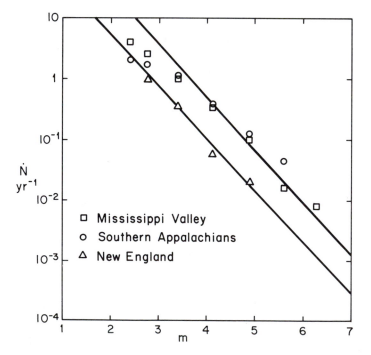

Figure 4.3. Number of earthquakes per year $\dot{N}$ occurring in the eastern United States with magnitudes greater than m as a function of m (Chinnery, 1979). The lines represent (4.1) with $b = 0.85$.

estimate the magnitude of these earthquakes but they were near magnitude eight. The southern Appalachian region includes the site of the Charleston eqarthquake of 1886. This earthquake did considerable damage and is estimated to have had a magnitude of 7.5. A sizeable earthquake occurred near Boston in 1755; this is in the New England region but little information is available on its magnitude. Since the data in Figure 4.3 are based on recorded damage and felt intensities, they are not as reliable as the data in Figures 4.1 and 4.2, which are based on instrumental observations. Nevertheless, a comparison of Figure 4.3 with Figure 4.2 indicates that the probability of an earthquake occurring in the eastern United States is about 1/100 of the probability that the same sized earthquake will occur in southern California. Taking the New Madrid, Missouri earthquakes to have an equivalent magnitude $m = 8$ we can estimate the recurrence time from the data in Figure 4.3 to be about 5000 years.

We now return to our discussion of seismicity in southern California. The data given in Figure 4.2 can be used to predict the regional strain using (4.16). Substituting $\mu = 3 \times 10^{10}$ Pa, $b = 0.89$, $c = 1.5$, $d = 9.1$, $v = 48$ mm yr^{-1}, $m_{max} = 8.05$, and $\dot{a} = 1.4 \times 10^5$ yr^{-1} we find from (4.16) that $A_p = 1.5 \times 10^4$ km^2. Taking the depth of the seismogenic zone to be 15 km, the length of the seismogenic zone corresponding to this area is 1000 km. This is about a factor of two larger than the actual length of the San Andreas fault in southern California. This is reasonably good agreement considering the uncertainties in the parameters. However, there are two other factors that can contribute to this discrepancy.

(1) Southern California is an area of active compressional tectonics. The Transverse Ranges are in this region and are associated with the displacements on the San Andreas fault. The strains associated with the formation of the Transverse Range should be added to the strains associated with strike–slip displacements on the San Andreas fault system.

(2) South of San Bernardino displacements on the San Andreas fault are associated with small and moderate earthquakes; no great earthquakes are believed to occur on this section. With a maximum magnitude of about seven the expected level of seismicity would be about a factor of five greater than with a maximum magnitude eight earthquake. Thus a higher level of

seismicity on this section of the San Andreas could contribute to the high observed level.

Although there is certainly a significant range of errors, the results given above indicate that the measured frequency–magnitude statistics associated with the Gutenberg–Richter frequency–magnitude relation (4.1) can be used to assess seismic hazards. The regional $b(D)$ and $\dot{a}$ values can be used to estimate recurrence times for earthquakes of various magnitudes.

There are two end-member models that explain the fractal distribution of earthquakes. The first is that there is a fractal distribution of faults and each fault has its own characteristic earthquake. The second is that each fault has a fractal distribution of earthquakes. Observations strongly favor the first hypothesis. On the northern and southern locked sections of the San Andreas fault there is no evidence for a fractal distribution of earthquakes. Great earthquakes and their associated aftershock sequences occur, but between great earthquakes seismicity is essentially confined to secondary faults. A similar statement can be made about the Parkfield section of the San Andreas fault where moderate-sized earthquakes occurred in 1881, 1901, 1924, 1934, and 1966. There is no evidence for a fractal distribution of events on this section of the San Andreas fault. We therefore conclude that a reasonable working hypothesis is that each fault has a characteristic earthquake and a fractal distribution of earthquakes implies a fractal distribution of faults.

Although we can conclude that the frequency–size distribution of faults is fractal, the fractal dimension is not necessarily the same as that for earthquakes. Equal fractal dimensions would imply that the interval of time between earthquakes is independent of scale. This need not be the case.

Tectonic models for a fractal distribution of faults have been proposed by King (1983, 1986), Turcotte (1986b), King *et al.* (1988), and Hirata (1989a). Fractal distributions of faults that give well defined *b*-values have been proposed by Huang and Turcotte (1988) and Hirata (1989b). Hirata *et al.* (1987) found a fractal distribution of microfractures in laboratory experiments that stressed unfractured granite.

It is generally difficult to quantify the frquency–size distributions of faults. This is because the surface exposure is generally limited. Many faults are not recognized until earthquakes occur on them. A systematic study of the statistics of exposed joints and fractures has been given by Barton and Hsieh (1989). Basement rock near Yucca Mountain, Nevada was cleared of soil and the distribution of fractures mapped. Basement exposures are known as pavements and the results for their pavement 1000 are given in Figure 4.4. This exposure was

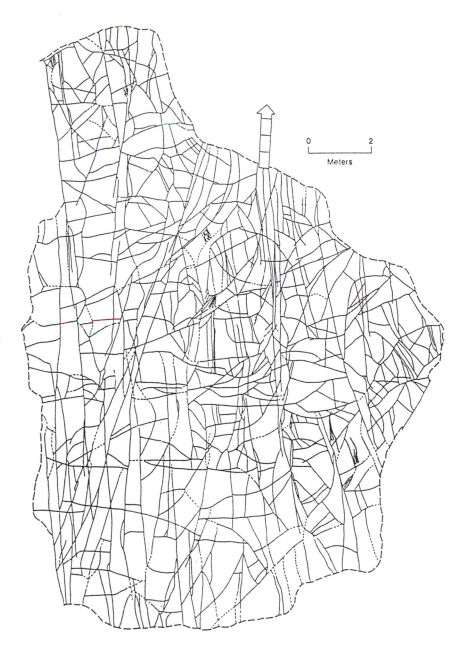

Figure 4.4. Map of the fractures and joints exposed on pavement 1000, Yucca Mountain, Nevada (Barton and Hsieh, 1989). This exposure was located in the densely welded orange brick unit of the Topopah Spring Member of the Miocene Paintbrush Tuff.

located in the densely welded orange brick unit of the Topopah Spring Member of the Miocene Paintbrush Tuff. These authors used the box-counting algorithm illustrated in Figure 2.8 to determine the fractal dimension of the exposure. Their result for the pavement illustrated in Figure 4.4 is given in Figure 4.5; the mean fractal dimension is $D = 1.7$.

The two-dimensional exposure can be used to infer the three-dimensional distribution of fracture (fault) sizes. We hypothesize that the model for fragmentation illustrated in Figure 3.3 is also applicable to tectonic fragmentation. A surface projection of this structure is given in Figure 4.6. Using the box-counting method on this fractal construction we have $N_1 = 1$ for $r_1 = h/2$ and $N_2 = 3$ for $r_2 = h/4$. From (2.2) we find that $D = \ln 3/\ln 2 = 1.5850$. Noting that

$$\frac{\ln 3}{\ln 2} + 1 = \frac{\ln 3 + \ln 2}{\ln 2} = \frac{\ln 6}{\ln 2} \tag{4.17}$$

we see that the fractal dimension of the cross section, $D_2 = \ln 3/\ln 2$,

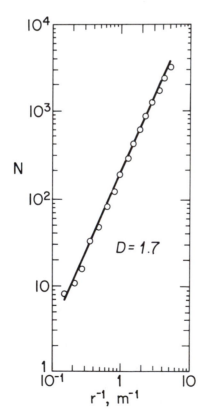

Figure 4.5. Statistics using the box-counting algorithm on the exposed fracture network at Yucca Mountain, Nevada (Barton and Hsieh, 1989) given in Figure 4.4. A correlation with (2.1) is used to obtain the fractal dimension $D = 1.7$.

Figure 4.6. Illustration of the surface exposure of the fractal fragmentation model given in Figure 3.3. With $r_1 = h/2$, $N_1 = 1$, and $r_2 = h/4$, $N_2 = 3$ we have $D = \ln 3/\ln 2 = 1.5850$.

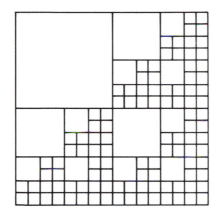

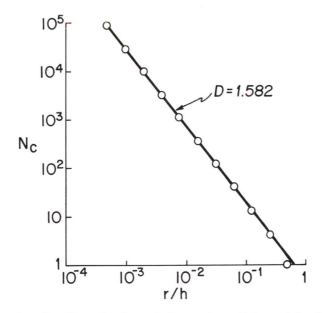

Figure 4.7. Cumulative statistics for the surface fracture model given in Figure 4.6. Correlation with (2.6) gives $D = 1.582$.

is related to the fractal dimension of the original construction from Figure 3.3, $D_3 = \ln 6/\ln 2$, by

$$D_3 = D_2 + 1 \qquad (4.18)$$

However, for a statistical model we again require cumulative statistics. From Figure 4.6 we have $N_{1c} = 1$ for $r_1 = h/2$, $N_{2c} = 4$ for $r_2 = h/4$, $N_{3c} = 13$ for $r_3 = h/8$, etc.; the cumulative statistics are given in Figure 4.7. An excellent fit with (2.6) is obtained for $D = 1.582$; (4.18) is a good approximation. The equivalent exposure from the comminution model is in good agreement with the observed results given in Figure 4.5.

Taking the fractal dimension for the distribution of faults to be $D_f = 2.6$, the number of faults with a characteristic linear dimension greater than r, in a given area, scales with r according to

$$N_f \sim \frac{1}{r^{2.6}} \qquad (4.19)$$

Similarly we assume that for earthquakes $D_e = 2$ so that the number of earthquakes per unit time, in a given area, with a characteristic rupture size greater than r scales with r according to

$$\dot{N}_e \sim \frac{1}{r^2} \qquad (4.20)$$

The average interval between earthquakes τ_e on a fault with a characteristic dimension r is given by

$$\tau_e = \frac{N_f}{\dot{N}_e} \sim \frac{r^2}{r^{2.6}} \sim \frac{1}{r^{0.6}} \qquad (4.21)$$

Thus the interval between earthquakes on a specified fault is longer for smaller faults. This is generally consistent with observations. If we further assume that faults remain active for a time τ_t then the total displacement δ on a fault of scale r is given by

$$\delta = \frac{\tau_t \delta_e}{\tau_e} \qquad (4.22)$$

where δ_e is the displacement in a single event. However

$$\delta_e \sim r \qquad (4.23)$$

and substitution of (4.21) and (4.23) into (4.22) gives

$$\delta \sim \tau_t r^{1.6} \qquad (4.24)$$

Walsh and Wattersen (1988) have correlated total fault displacement δ with fault length r for a variety of faults; their results are given in Figure 4.8. A good correlation with (4.24) is found. It should be emphasized that this correlation must be to some extent fortuitous since τ_t is unlikely to be a constant in different tectonic settings.

The fractal distribution of earthquake magnitudes is a universal feature of distributed seismicity on a variety of scales. There is extensive evidence that this result implies a fractal distribution of fault sizes. Although major plate boundaries are characterized by a dominant fault such as the San Andreas, the evolution of plates is not consistent with displacements on a single fault (Dewey, 1975). Thus regional deformation must occur. The hypothesis that this deformation follows the comminution model provides a fractal distribution with a fractal dimension consistent with observations.

We next turn to volcanic eruptions. It is considerably more difficult to quantify a volcanic eruption than it is to quantify an earthquake. There are a variety of types of eruption and the various types are quantified in different ways. Some eruptions produce primarily magma (liquid rock) while others produce primarily tephra (ash). Utilizing the volume of tephra as a measure of size McClelland *et al.* (1989) have published frequency–volume statistics for volcanic eruptions. Their results for eruptions during the period 1975 to 1985 and for historic (last 200 years) eruptions are given in Figure 4.9. The number of eruptions with a volume of tephra greater than a specified value is given as a function of the volume. A reasonably good correlation is obtained with the fractal relation (2.6) by taking $D = 2.14$. It appears that volcanic eruptions are scale invariant over a significant range of sizes.

A single volcano can produce eruptions with a wide spectrum of sizes. Also, volcanoes have a wide spectrum of sizes. The circumstances that determine the volume of tephra in an eruption are poorly

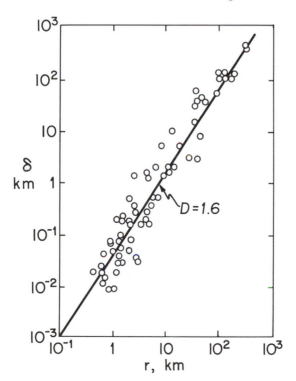

Figure 4.8. Dependence of total fault displacement on fault length; data from Walsh and Wattersen (1988).

understood. Thus models that would provide an explanation of the observed value of D are not available.

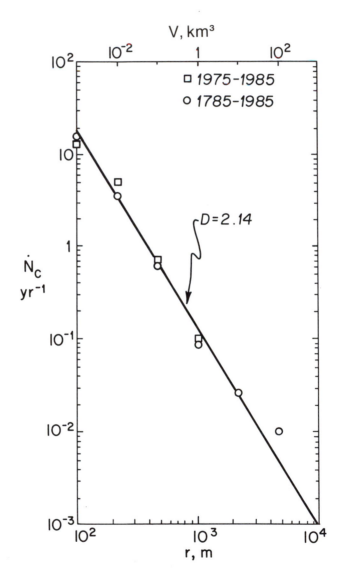

Figure 4.9. Number of volcanic eruptions per year $\dot{N}_c$ with a tephra volume greater than V as a function of V for the period 1975–1985 (squares) and for the last 200 years (circles) (McClelland *et al.*, 1989). The line represents the correlation with (2.6) taking $D = 2.14$.

Problems

Problem 4.1. Determine E_s, M, A, and δ_e for an $m = 7$ earthquake (take $c = 1.5$, $d = 9.1$, $\alpha = 3.27 \times 10^6$ Pa, $\mu = 3 \times 10^{10}$ Pa).

Problem 4.2. Determine E_s, M, A, and δ_e for an $m = 6$ earthquake (take $c = 1.5$, $d = 9.1$, $\alpha = 3.27 \times 10^6$ Pa, $\mu = 3 \times 10^{10}$ Pa).

Problem 4.3. On a world-wide basis how many magnitude six earthquakes are expected in a year?

Problem 4.4. In a region the recurrence interval τ_e for a magnitude six earthquake is 18 months; if $b = 0.9$ what is the recurrence interval τ_e for a magnitude seven earthquake?

Problem 4.5. In a region the recurrence interval τ_e for a magnitude five earthquake is 10 years; if $b = 1$ what is the recurrence interval τ_e for a magnitude seven earthquake?

Problem 4.6. In a subduction zone the length of the seismogenic zone is 1000 km and its depth is 30 km. The convergence velocity is 100 mm yr^{-1}.
(a) Determine $\dot{a}$ if $b = 1$, $c = 1.5$, $d = 9.1$, $\mu = 3 \times 10^{10}$ Pa, and $m_{max} = 8.5$.
(b) Determine the recurrence time for the magnitude 8.5 earthquake.

Problem 4.7. The length of a seismogenic zone on a strike–slip fault is 100 km and its depth is 15 km. (a) Determine $\dot{a}$ if $b = 1$, $c = 1.5$, $d = 9.1$, $\mu = 3 \times 10^{10}$ Pa, $v = 50$ mm yr^{-1}, and $m_{max} = 6.2$. (b) Determine the recurrence time for the magnitude 6.2 earthquake.

Problem 4.8. The charactertistic earthquake of magnitude seven on a fault has a recurrence interval of 200 years; using (4.21), what is the recurrence time for a characteristic earthquake of magnitude six? Take $c = 1.5$, $d = 9.1$.

Problem 4.9. The characteristic earthquake of magnitude six on a fault has a recurrence interval of 400 years; using (4.21), what is the recurrence time for a characteristic earthquake of magnitude four? Take $c = 1.5$ and $d = 9.1$.

Problem 4.10. Ash eruptions with a volume greater than 1000 km^3 are expected to have a profound influence on the global climate. What is the expected recurrence interval for such eruptions?

CHAPTER FIVE

Ore grade and tonnage

Statistical treatments of ore grade and tonnage for economic ore deposits have provided a basis for estimating ore reserves. The objective is to determine the tonnage of ore with grades above a specified value. The grade is defined as the ratio of the mass of the mineral extracted to the mass of the ore. Evaluations can be made either on a global or a regional basis. Much of the original work on this problem was carried out by Lasky (1950). He argued that a linear relation is obtained if the logarithm of the tonnage of ore with grades above a specified value is plotted against grade.

Other authors, however, have suggested that a linear relation is obtained if the logarithm of the tonnage of ore with grades above a specified value is plotted against the logarithm of the grade. The latter is a fractal relation. A fractal relation would be expected if the concentration mechanism is scale invariant. Many different mechanisms are responsible for the concentrations of mineral that lead to economic ore deposits. Probably the most widely applicable mechanisms are associated with hydrothermal circulations. However, even the most thoroughly studied mechanisms are not sufficiently well understood to allow the development of quantitative models of ore concentration.

We first consider two simple models that illustrate the log-normal and power-law distributions for tonnage versus grade. De Wijs (1951, 1953) proposed the model for mineral concentration that is illustrated in Figure 5.1(a). In this model an original mass of rock M_0 is divided into two equal parts each with a mass $M_1 = M_0/2$. The original mass of the rock has a mean mineral concentration C_0, which is the ratio of the mass of mineral to mass of rock. As in Chapter 3 we refer to this mass as a zero-order cell. It is hypothesized that the mineral is concentrated into one of the two zero-order elements such that one element is enriched and the other element is depleted. The zero-order elements then become first-order cells each of which is divided into two first-order elements with mass $M_2 = M_0/4$.

The mean mineral concentration in the enriched zero-order element C_{11} is given by

$$C_{11} = \phi_2 C_0 \tag{5.1}$$

where ϕ_2 is the enrichment factor. The first subscript on C refers to the order of cell being considered. The second subscript refers to the amount of enrichment: the lower the number the more the enrichment and the higher the concentration. The subscript on the enrichment factor refers to the fact that each cell is divided into two equal

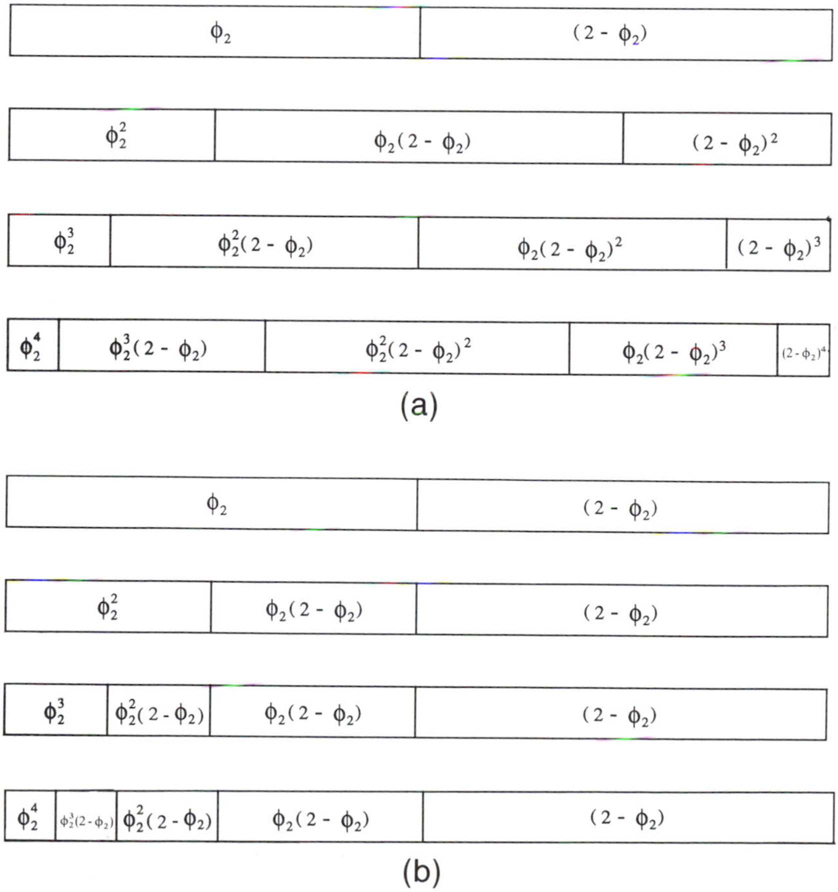

Figure 5.1. Illustration of two models for the concentration of economic ore deposits. In both models a mass of rock is first divided into two equal parts, then four equal parts, etc. (a) De Wijs (1951, 1953) proposed a successive concentration of minerals into smaller and smaller volumes. Four orders of concentration are illustrated. At each order one-half of each cell is enriched by the ratio ϕ_2 and the other half is depleted by the ratio $2 - \phi_2$. This model gives a binominal distribution for tonnage versus grade and in the limit of very small volumes gives a log-normal relation. (b) Turcotte (1986c) proposed a similar model, but with the further concentration limited to the highest-grade ores. This leads to a power-law (fractal) distribution of tonnage versus grade.

(a)

(b)

elements; the enrichment factor ϕ_2 is greater than unity since C_{11} must be greater than C_0. A simple mass balance shows that the concentration in the depleted zero-order element is

$$C_{12} = (2 - \phi_2)C_0 \tag{5.2}$$

The enrichment factor must be in the range $1 < \phi_2 < 2$. This model is illustrated in Figure 5.1(a). The process of concentration is then repeated at the next order as illustrated in Figure 5.1(a). The zero-order elements become first-order cells and each cell is again divided into two elements of equal mass $M_2 = M_0/4$. The mineral is again concentrated by the same ratio into each first-order element. The enriched first-order element in the enriched first-order cell has a concentration

$$C_{21} = \phi_2^2 C_0 \tag{5.3}$$

The depleted first-order element of the enriched first-order cell and the enriched first-order element of the delpeted first-order cell both have the same concentrations:

$$C_{22} = C_{23} = \phi_2(2 - \phi_2)C_0 \tag{5.4}$$

The depleted first-order element of the depleted first-order cell has a concentration

$$C_{24} = (2 - \phi_2)^2 C_0 \tag{5.5}$$

This result is also illustrated in Figure 5.1(a) along with two higher-order cells. This model gives a binomial distribution of ore grades and in the limit of infinite order reduces to the log-normal distribution given in (3.14). The resulting distribution is not scale invariant; the reason is that the results are dependent on the size of the initial mass of ore chosen and this mass enters into the tonnage–grade relation.

Cargill *et al.* (1980, 1981) disagreed with the logarithmic dependence and suggested that a linear relationship is obtained if the logarithm of the tonnage is plotted against the logarithm of the mean grade. A simple model that gives this result was proposed by Turcotte (1986c) and is illustrated in Figure 5.1(b). This model follows very closely the model discussed above. Again, an original mass of rock M_0 is divided into two parts each with a mass $M_1 = M_0/2$ and it is hypothesized that the mineral is concentrated into one of the two zero-order elements so that (5.1) and (5.2) are applicable. However, at the next step only the enriched element is further fractionated.

The problem is renormalized so that the enriched element is treated in exactly the same way at every scale (order). This results in a fractal (scale-invariant) distribution. The concentration of ore into one of the two elements in the enriched first-order cell results in the concentrations given by (5.3) and (5.4). However, the depleted first-order cell continues to have the concentration given by (5.2). This result is illustrated in Figure 5.1(b) along with two higher-order cells.

The results given in Figure 5.1(b) can be generalized to the nth order with the result

$$C_n = \phi_2^n C_0 \tag{5.6}$$

where C_n is the mean ore grade associated with the mass

$$M_n = \frac{1}{2^n} M_0 \tag{5.7}$$

Taking the natural logarithms of (5.6) and (5.7) gives

$$\ln\left(\frac{C_n}{C_0}\right) = n \ln \phi_2 \tag{5.8}$$

and

$$\ln\left(\frac{M_n}{M_0}\right) = -n \ln 2 \tag{5.9}$$

The elimination of n between (5.8) and (5.9) gives

$$\ln\left(\frac{C_n}{C_0}\right) = -\frac{\ln \phi_2}{\ln 2} \ln\left(\frac{M_n}{M_0}\right) = \frac{\ln \phi_2}{\ln 2} \ln\left(\frac{M_0}{M_n}\right) \tag{5.10}$$

With the density assumed to be constant, $M \sim r^3$, where r is the linear dimension of the ore deposit considered, and we have

$$\frac{C_n}{C_0} = \left(\frac{M_0}{M_n}\right)^{\frac{\ln \phi_2}{\ln 2}} = \left(\frac{r_0}{r_n}\right)^{\frac{3\ln \phi_2}{\ln 2}} \tag{5.11}$$

Comparison with (2.6) shows that this is a power-law or fractal distribution with

$$D = 3 \frac{\ln \phi_2}{\ln 2} \tag{5.12}$$

Since the allowed range for ϕ_2 is $1 < \phi_2 < 2$, the allowed range for the fractal dimension is $0 < D < 3$. In order to be fractal the distribution must be scale invariant. The scale invariance is clearly illustrated in Figure 5.1(b). The concentration of ore could be started at any order and the same result would be obtained. The left half

at order two looks like order one, the left half at order three looks
like order two, etc. This is not true for the distribution illustrated in
Figure 5.1(a).

The value of ϕ is dependent on the ratio of the mass of the enriched
element to the mass of the depleted element. However, we now show
that the derived fractal relation is independent of this choice. We
apply our scale-invariant model for ore concentration to the cubic
model for fragmentation illustrated in Figure 3.2. We consider a
zero-order cell that is a cube with linear dimension h. This cell is
divided into eight zero-order cubic elements, each with linear
dimension $h/2$. We hypothesize that a fraction of the mineral in a
cell is concentrated into one element. If the mean concentration of
the mineral in the zero-order cell is C_0 and in the zero-order element
is C_1 we define a concentration ratio

$$\phi_8 = \frac{C_1}{C_0} \tag{5.13}$$

The subscript eight is written because we now consider the case in
which the mass of each element, M_1, is related to the mass of the
cell, M_0, by

$$\frac{M_1}{M_0} = \frac{1}{8} \tag{5.14}$$

The zero-order element into which the mineral is concentrated
becomes a first-order cell; this cell is divided into eight cubic
first-order elements each with linear dimension $h/4$. The process of
mineral concentration is repeated so that

$$C_2 = \phi_8 C_1 = \phi_8^2 C_0 \tag{5.15}$$

and

$$M_2 = \frac{1}{8}M_1 = \frac{1}{8^2}M_0 \tag{5.16}$$

The above results are generalized to the nth order to give

$$C_n = \phi_8^n C_0 \tag{5.17}$$

and

$$M_n = \frac{1}{8^n}M_0 \tag{5.18}$$

Taking the natural logarithms of (5.17) and (5.18) yields

$$\ln\left(\frac{C_n}{C_0}\right) = n \ln \phi_8 \tag{5.19}$$

and

$$\ln\left(\frac{M_n}{M_0}\right) = -n \ln 8 \tag{5.20}$$

Elimination of n between (5.19) and (5.20) gives

$$\ln\left(\frac{C_n}{C_0}\right) = -\frac{\ln \phi_8}{\ln 8} \ln\left(\frac{M_n}{M_0}\right) = \frac{\ln \phi_8}{\ln 8} \ln\left(\frac{M_0}{M_n}\right) \tag{5.21}$$

or

$$\frac{C_n}{C_0} = \left(\frac{M_0}{M_n}\right)^{\frac{\ln \phi_8}{\ln 8}} = \left(\frac{r_0}{r_n}\right)^{\frac{3 \ln \phi_8}{\ln 8}} \tag{5.22}$$

Comparison with (2.6) shows that this is a power-law or fractal distribution with

$$D = 3\frac{\ln \phi_8}{\ln 8} \tag{5.23}$$

We next show that (5.23) is entirely equivalent to (5.12). The first-order concentration into one-eighth of the original mass, ϕ_8, must be equivalent to three orders of the concentration into one-half the original mass, ϕ_2. Thus we can write

$$\phi_8 = \phi_2^3 \tag{5.24}$$

It follows that

$$\frac{\ln \phi_8}{\ln 8} = \frac{\ln \phi_2^3}{\ln 8} = \frac{3 \ln \phi_2}{3 \ln 2} = \frac{\ln \phi_2}{\ln 2} \tag{5.25}$$

Thus (5.23) is equivalent to (5.12) and can be derived independently of the mass ratio chosen.

In each of the enrichment steps in our fractal model the concentration C_n is the mean concentration in the mass of ore M_n. For applications to actual mineral deposits we generalize the fractal relation between ore grade and tonnage to

$$\frac{\bar{C}}{C_0} = \left(\frac{M_0}{M}\right)^{D/3} \tag{5.26}$$

where M is the mass of the highest grade ores, which have a mean concentration $\bar{C}$. The reference mass M_0 is the mass of rock from which the ore was derived, which has a mean concentration C_0.

As in the previous examples of naturally occurring fractal

distributions, there are limits to the applicability of (5.26). The lower limit on the ore grade is clearly the regional background grade C_0 that has been concentrated. However, there is also an upper limit: the grade $\bar{C}$ cannot exceed unity, which corresponds to pure mineral.

The entire subject of tonnage–grade relations has been reviewed by Harris (1984). There is clearly a controversy in the literature between Lasky's law, which gives an exponential dependence of tonnage on grade, and the power-law or fractal dependence. Lasky (1950) and Musgrove (1965) have argued in favor of the exponential relation. On the other hand, Cargill *et al.* (1980, 1981) have argued in favor of the power-law dependence. These authors based their analyses on records of annual production and mean grade. Their results for mercury production in the United States are given in Figure 5.2. The cumulative tonnage of mercury mined prior to a specified date is divided by the cumulative tonnage of ore from which the mercury was obtained to give the cumulative average grade. The data points in Figure 5.2 represent the five-year cumulative average grade (in weight ratio) versus the cumulative tonnage of ore. Using Bureau of Mines records Cargill *et al.* found that the total amount of mercury mined between 1890 and 1895 was M_{m1} and the tonnage of ore from which this mercury was obtained was M_1; the mean grade for this period was $\bar{C}_1 = M_{m1}/M_1$. The cumulative amount of mercury mined between 1890 and 1900 was M_{m2} and the cumulative

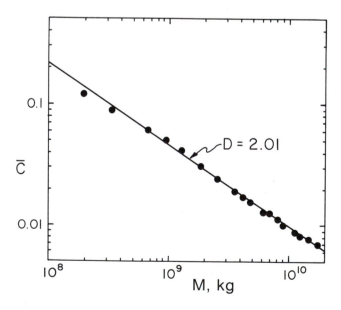

Figure 5.2. Dependence of cumulative ore tonnage M on mean grade $\bar{C}$ for mercury production in the United States (Cargill *et al.*, 1981). Correlation with (5.26) gives a fractal dimension $D = 2.01$.

tonnage of ore from which the mercury was mined was M_2; the mean cumulative grade for this period was $\bar{C}_2 = M_{m^2}/M_2$. These computations represent the two data points furthest to the left in Figure 5.2. The other data points represent the inclusion of additional five-year periods in the computations. Cargill *et al.* (1980, 1981) further hypothesized that the highest-grade ores are usually mined first so that the cumulative ratio of mineral tonnage to ore tonnage at a given time is a good approximation to the mean ore grade of the highest grade ores. Thus it is appropriate to compare their data directly with the fractal relation (5.26). Excellent agreement is obtained taking $D = 2.01$. This is strong evidence that the enrichment processes leading up to the formation of mercury deposits are scale invariant.

It is also of interest to introduce a reference concentration of mercury into the fractal relation. An appropriate choice is the mean measured concentration in the continental crust. The mean crustal concentration of mercury as given by Taylor (1964) is $\bar{C}_0 = 8 \times 10^{-8}$ (0.08 ppm). Using this value in (5.26) we find that the correlation line in Figure 5.2 is given by

$$M = 4.05 \times 10^{17} \left(\frac{8 \times 10^{-8}}{\bar{C}} \right)^{1.49} \tag{5.27}$$

with M in kilograms. According to the fractal model the mercury ore in the United States has been concentrated from continental crust with a mass $M_0 = 4.05 \times 10^{17}$ kg. Assuming a mean crustal density of $2.7 \times 10^3 \, \text{kg m}^{-3}$, the mercury resources of the United States were concentrated from an original crustal volume of $1.5 \times 10^5 \, \text{km}^3$. Since the total crustal volume of the United States is approximately $2.7 \times 10^8 \, \text{km}^3$, the source volume for the mercury deposits is about 0.05 per cent of the total. It is concluded that the processes responsible for the enrichment of mercury ore deposits are restricted to a relatively small fraction of the crustal volume.

It is seen from Figure 5.2 that the cumulative production of 1.2×10^8 kg of mercury has been obtained from 2×10^{10} kg of ore of volume $7.4 \times 10^6 \, \text{m}^3$. Since the source region has a volume of $1.5 \times 10^5 \, \text{km}^3$, the fraction of the source region that has been mined is only 5×10^{-8}. The results given in Figure 5.2 can also be used to determine how much mercury ore must be mined in the future in order to produce a specified amount of mercury. To produce the next 1.2×10^8 kg of mercury will require the processing of about 1.6×10^{11} kg of ore.

Using copper production records in the same way, Cargill *et al.* (1981) have also given cumulative grade–tonnage data for copper production in the United States. Their results are given in Figure 5.3. The cumulative grade is again given as a function of cumulative ore tonnage at five-year intervals. The data obtained prior to 1920 fall systematically low compared to the later data. Cargill *et al.* (1981) attributed this systemic deviation from a fractal correlation to the adoption of an improved metallurgical technology for the extraction of copper in the 1920s. A smaller fraction of the available copper was extracted prior to this time so that the data points are low. It is again appropriate to compare this data with the fractal relation (5.26). Assuming the early data to be systematically low, excellent agreement is obtained taking $D = 1.16$. This fractal dimension is almost a factor of two less than the fractal dimension obtained for mercury ore. This indicates that the applicable enrichment processes concentrate copper less strongly than they do mercury.

We again relate the fractal relation for the enrichment to the mean crustal concentration. The mean concentration of copper in the upper crust as given by Taylor and McLennan (1981) is $\bar{C}_0 = 2.5 \times 10^{-5}$ (25 ppm). Using this value in (5.26), we find that the correlation line in Figure 5.3 is given by

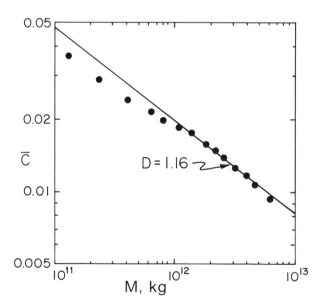

Figure 5.3. Dependence of cumulative ore tonnage M on mean grade $\bar{C}$ for copper production in the United States (Cargill *et al.*, 1981). Correlation with (5.26) gives a fractal dimension $D = 1.16$.

$$M = 3.22 \times 10^{19} \left(\frac{2.5 \times 10^{-5}}{\bar{C}} \right)^{2.59} \tag{5.28}$$

with M in kilograms. According to the fractal model the copper ore in the United States has been concentrated from continental crust with a mass $M_0 = 3.22 \times 10^{19}$ kg. Assuming a mean upper crustal density of 2.7×10^3 kg m^{-3}, the copper resources of the United States were concentrated from an original crustal volume of 1.19×10^7 km^3. This represents about 4 per cent of the total crustal volume of the United States. The crustal volume from which copper is enriched is nearly 100 times larger than the volume from which mercury is enriched. It is concluded that the processes responsible for the enrichment of copper are much more widely applicable than those for mercury.

As our final example we consider data on the relationship between cumulative tonnage and grade for uranium in the United States. Data for the preproduction inventory as given by the US Department of Energy has been tabulated by Harris (1984, p. 228) in terms of cumulative tonnage and the average grade of this tonnage; this data is tabulated in Figure 5.4. The high-grade data is based on production records and the lower-grade data is based on estimates of reserves. The higher-grade data are in excellent agreement with the fractal

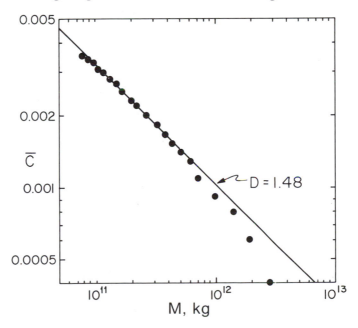

Figure 5.4. Dependence of cumulative ore tonnage M on mean grade $\bar{C}$ for uranium reserves in the United States (Harris, 1984, p. 228). Correlation with (5.26) gives a fractal dimension $D = 1.48$.

relation (5.26) taking $D = 1.48$. Thus the enrichment of uranium is intermediate between the enrichment of copper and mercury. The predicted cumulative tonnage at lower grades falls below the extrapolation of the fractal relation; this can be attributed to an underestimation of the preproduction inventory at low grades.

It is again instructive to relate the fractal relation for the enrichment of uranium to the mean crustal concentration. The mean concentration of uranium in the upper crust as given by Taylor and McLennan (1981) is $\bar{C}_0 = 1.25 \times 10^{-6}$ (1.25 ppm). Using this value in (5.26), we find that the correlation line in Figure 5.4 is given by

$$M = 6.4 \times 10^{17} \left(\frac{1.25 \times 10^{-6}}{\bar{C}} \right)^{2.03} \tag{5.29}$$

with M in kilograms. According to the fractal model the uranium ore in the United States has been concentrated from continental crust with a mass $M_0 = 6.4 \times 10^{17}$ kg. Assuming a mean crustal density of 2.7×10^3 km m^{-3}, the uranium resources of the United States were concentrated from an original crustal volume of 2.4×10^5 km^3. This represents about 0.09 percent of the crustal volume of the United States. The crustal volume from which uranium is enriched is about a factor of two larger than the crustal volume for mercury but is a factor of fifty less than the crustal volume for copper.

In several examples the statistics on ore tonnage versus ore grade have been shown to be fractal to a good approximation. This is indicative that the physical and chemical processes that lead to the formation of economic ore deposits are scale invariant over a reasonably wide range of scales. This is not really surprising, but does not constrain alternative models for ore formation. The examples given yield a considerable range of fractal dimensions. The fractal dimensions found were 2.01 for mercury and 1.16 for copper. This may indicate that different scale-invariant processes are responsible for the enrichment of different minerals.

There is also evidence that the frequency–size distribution of oil fields obeys fractal statistics. Drew *et al.* (1982) used the relation $N_{i-1} = 1.67 N_i$ in order to estimate the number of fields of order i, N_i, in the western Gulf of Mexico. Since the volume of oil in a field of order i is a factor of two greater than the volume of oil in a field of order $i - 1$, their relation is equivalent to a fractal distribution with $D = 2.22$. The number–size statistics for oil fields in the United

States (excluding Alaska) as compiled by Ivanhoe (1976) are given in Figure 5.5. A reasonably good correlation with the fractal relation (2.6) is obtained taking $D = 4.16$. The large difference between these two values for the fractal dimension may be attributed to differences in the regional geology, but it may also be due to difficulties in the data. It is often difficult to determine whether adjacent fields are truly separate and data on reserves are often poorly constrained. Nevertheless, the applicability of fractal statistics to petroleum reserves can have important implications. Reserve estimates for petroleum have been obtained by using power-law (fractal) statistics and log-normal statistics. Accepting power-law statistics leads to considerably higher estimates for available reserves.

The model for the concentration of economic ore deposits given above leads to a range of geometrically acceptable fractal dimensions. However, the observed distribution for oil fields falls outside this range. This again illustrates the difficulties associated with restrictions on power-law (fractal) distributions. As stated previously, we define a power-law statistical distribution as a fractal distribution.

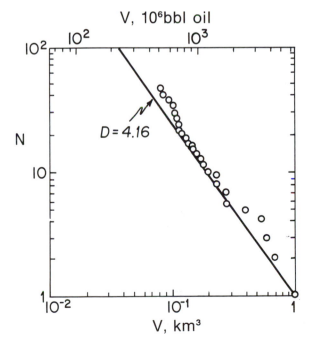

Figure 5.5. The number N of oil fields in the United States with a volume of oil greater than V as a function of V. The equivalent number of barrels is also given. The circles represent the data given by Ivanhoe (1976) and the line represents the correlation with (2.6) taking $D = 4.16$.

Problems

Problem 5.1. Determine the concentration factor ϕ_2 for an ore deposit with $D = 2$.

Problem 5.2. Determine the concentration factor ϕ_2 for an ore deposit with $D = 1$.

Problem 5.3. Consider the cubic model for mineral concentration illustrated in Figure 3.2. (a) In terms of the enrichment factor ϕ_8 defined by (5.1) and C_0, what is the concentration in the seven depleted zero-order elements? (b) What is the concentration in the seven depleted first-order elements? (c) What is the allowed range for ϕ_8? (d) What is the corresponding allowed range for D?

Problem 5.4. From the correlation for mercury production given in (5.27) determine the total production of mercury when the mean grade of ore that has been mined reaches $\bar{C} = 0.001$.

Problem 5.5. From the correlation for copper production given in (5.28) determine the total production of copper to date. Assume that the mean grade of ore mined prior to the present time is $\bar{C} = 0.008$.

Problem 5.6. Taking the correlation for uranium production given in (5.29) determine the total production of uranium to date. Assume that the mean grade of ore mined prior to the present time is $\bar{C} = 0.001$.

CHAPTER SIX

Fractal clustering

We next relate fractal distributions to probability. This can be done using the sequence of line segments illustrated in Figure 2.1. The objective is to determine the probability that a step of length r will include a line segment. First consider the construction illustrated in Figure 2.1(a). At zero order the probability that a step of length $r_0 = 1$ will encounter a line segment, $p_0 = 1$; at first order we have $r_1 = \frac{1}{2}$ and $p_1 = \frac{1}{2}$, and at second order $r_2 = \frac{1}{4}$ and $p_2 = \frac{1}{4}$. Next consider the construction illustrated in Figure 2.1(c). At zero order the probability that a step of length $r_0 = 1$ will encounter a line segment is $p_0 = 1$; at first order we have $r_1 = \frac{1}{2}$ and $p_1 = 1$, and at second order $r_2 = \frac{1}{4}$ and $p_2 = 1$. Finally we consider the Cantor set illustrated in Figure 2.1(e). At zero order the probability that a set of length $r_0 = 1$ will encounter a line segment is $p_0 = 1$; at first order we have $r_1 = \frac{1}{3}$ and $p_1 = \frac{2}{3}$, and at second order $r_2 = \frac{1}{9}$ and $p_2 = \frac{4}{9}$.

The probability that a step of length r_i will include a line segment can be generalized to

$$p_n = N_n r_n \tag{6.1}$$

where N_n is the number of line segments of length r_n. Taking $C = 1$ in (2.1) the number N_n is related to r_n so that we obtain

$$p_n = r_n^{1-D} \tag{6.2}$$

For the Cantor set the probability that a step of length $r_n = (\frac{1}{3})^n$ encounters a line segment is $p_n = (\frac{2}{3})^n$ so that $D = \ln 2/\ln 3$ as was obtained previously.

The Cantor set is both scale invariant and deterministic. Its deterministic aspect can be eliminated quite easily. A scale-invariant random set is generated by randomly removing one-third of each line rather than always removing the middle third. This process is illustrated in Figure 6.1. The fractal dimension is unchanged and the probability relations derived above are still applicable.

We will use the examples given above as the basis for studying fractal clustering. We consider a series of point events that occur at specified times. In order to consider N point events that have occurred in the time interval τ_0 we introduce the natural period $\tau_N = \tau_0/N$. We then introduce a sequence of intervals defined by

$$\tau_n = \frac{\tau_0}{n}, \, n = 1, 2, \ldots, N \tag{6.3}$$

Our measure of clustering will be the probability p_n that an event occurs in an interval of length τ_n.

As a specific example, consider a uniform (equally spaced in time) series of N events that occur in an interval τ_0. The first event occurs at $t = \tau_0/2N$, the second event at $t = 3\tau_0/2N$, the third event at $t = 5\tau_0/2N$, and so forth. The probability p_n that an event will occur in an interval is given by

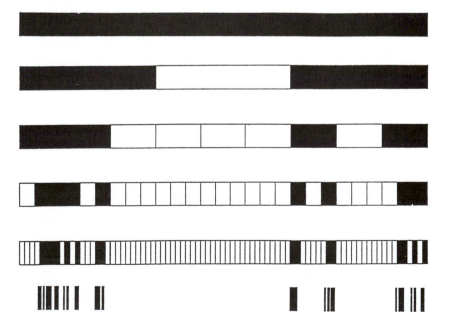

Figure 6.1. Illustration of the first six orders of a random Cantor set. At each step, a random third of each solid line is removed.

$$p_n = \begin{cases} \dfrac{\tau_n N}{\tau_0} = \dfrac{N}{n} & \text{if } \tau_n < \dfrac{\tau_0}{N} \\[3mm] 1 & \text{if } \tau_n > \dfrac{\tau_0}{N} \end{cases} \tag{6.4}$$

If the number of events N is greater than the number of intervals n we have $N > n$ or $\tau_n > \tau_0/N$. In this case an event occurs in every interval so that $p_n = 1$. If the number of events is less than the number of intervals we have $N < n$ or $\tau_n < \tau_0/N$. In this case only N of the n intervals have events so that $p_n = N/n$. Because there is no clustering, no interval τ_n contains more than one event.

A more realistic model for a series of events in time is that their occurrence is completely uncorrelated. The time at which each individual event occurs is random. An example would be telephone calls placed in a city during a given hour. If N point events occur randomly in a time interval τ_0 it is a Poisson distribution. In the limit of a very large number of events ($N \to \infty$) the distribution of intervals between events is given by

$$f(\tau) = \frac{N}{\tau_0} \exp\left(-\frac{N\tau}{\tau_0}\right) \tag{6.5}$$

where $f(\tau)\,d\tau$ is the probability that an event will occur after an interval of time between τ and $\tau + d\tau$ in length. This distribution is clearly not scale invariant since the natural time scale τ_0/N enters (6.5).

We will next determine the probability p that an interval of length τ_n will include an event if N events occur randomly in an interval τ_0. This is the classic problem of the random distribution of N balls into n boxes. We introduce $P_m = m/n$ where m is the number of intervals that include events and $n = \tau_0/\tau_n$ is the total number of intervals. We assume both n and m are integers. The probability that P_m has a specified value is given by

$$f_m(P_m) = \left(\frac{n}{n(1 - P_m)}\right) \sum_{\mu=0}^{nP_m} (-1)^\mu \binom{nP_m}{\mu} \left(P_m - \frac{\mu}{n}\right)^N \tag{6.6}$$

where $\sum_{m=0}^{n} f_m(P_m) = 1$; the binomial coefficient is defined by

$$\binom{n}{m} = \frac{n!}{m!(n-m)!} \tag{6.7}$$

It is often appropriate to take N and n to be large numbers. In this limit, $f_n(P_m)$ from (6.6) has a strong maximum at a specific value of P_m, p.

In order to illustrate the probabilistic approach to clustering we consider a ninth-order random Cantor set. First we rescale so that unit length is the length of a ninth-order element. Thus the length of the zero-order element is 3^9 and there are 2^9 ninth-order elements. In order to determine the fractal dimension of the clustering by the 'box method', we take intervals of length $r_n = 2^n$ and determine the fraction p that include at least one ninth-order element as a function of r_n. An example is given by the open circles in Figure 6.2. The best-fit straight line has a slope of 0.368 so that $p \sim r^{0.368}$ and $D = 0.632$. The deviation from the exact value $D = 0.6309$ for the deterministic Cantor set is due to the reduced rate of curdling in the probabilistic set. If the same number of ninth-order elements is uniformly distributed (no clustering), the probability of finding an element with an interval from (6.4) is given by the solid circles in Figure 6.2. In this case, the slope is unity for $r < (\frac{3}{2})^9$ and zero for $r > (\frac{3}{2})^9$. Thus, $D = 0$ for $r < (\frac{3}{2})^9$, i.e. a set of isolated points, and $D = 1$ for $r > (\frac{3}{2})^9$, i.e. a line.

Fractal clustering has been applied to seismicity by Sadovskiy *et al.* (1985) and by Smalley *et al.* (1987). The latter authors considered the temporal variation of seismicity in several regions near Efate Island in the New Hebrides island arc for the period 1978 to 1984. One of their examples is given in Figure 6.3. During the period under

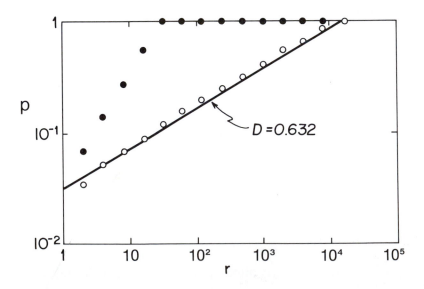

Figure 6.2. The fraction p of steps of length r that include solid lines for a ninth-order random Cantor set is given by the open circles. The unit length is the length of the shortest line; the original line length is 3^9. The solid circles correspond to a uniform distribution of the same number of lines as given by (6.4). The line corresponds to (6.2) with $D = 0.632$.

consideration 49 earthquakes that exceeded the minimum magnitude required for detection occurred in the region. Time intervals τ such that $8 \text{ min} \leqslant \tau \leqslant 524\,288 \text{ min}$ were considered. The fraction of intervals with earthquakes p as a function of interval length τ is given in Figure 6.3(a) as the open circles. The solid line shows the correlation with the fractal relation (6.2) with $D = 0.255$. The dashed line is the result for uniformly spaced events. The results of a simulation for a random distribution of 49 events in the time interval studied is given in Figure 6.3(b). The random simulation (Poisson distribution) is significantly different from the earthquake data and is close to the uniform distribution.

Fractal clustering can also be studied in higher dimensions. The application to two dimensions is illustrated by the sequence of constructions given in Figure 2.2. The objective is to determine the probability that a square box of size r encounters a square that has been retained. First consider the construction given in Figure 2.2(a). At zero order the probability that a box

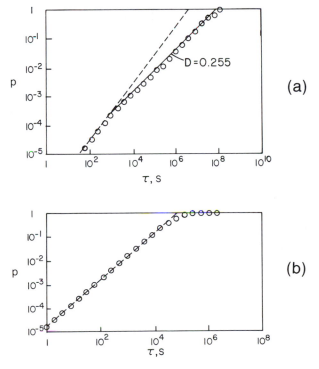

Figure 6.3. Fractal cluster analysis of 49 earthquakes that occurred near Efate Island, New Hebrides in the period 1978–1984 (Smalley *et al.*, 1987). (a) The circles give the fraction of intervals p of length τ that include an earthquake as a function of τ. The solid line represents the correlation with (6.2) taking $D = 0.255$. The broken line is the result for uniformly spaced events. (b) The results for 49 randomly distributed events (Poisson process).

of size $r_0 = 1$ will include a square is $p_0 = 1$; at first order we have $r_1 = \frac{1}{3}$ and $p_1 = \frac{1}{9}$, and at second order we have $r_2 = \frac{1}{9}$ and $p_2 = \frac{1}{81}$. Next consider the construction illustrated in Figure 2.2(b). At zero order the probability that a box of size $r_0 = 1$ will include a retained square is $p_0 = 1$; at first order we have $r_1 = \frac{1}{3}$ and $p_1 = \frac{2}{9}$, and at second order we have $r_2 = \frac{1}{9}$ and $p_2 = \frac{4}{81}$. Finally we consider the Sierpinski carpet illustrated in Figure 2.2(d). At zero order the probability that a box of size $r_i = 1$ will include a retained square is $p_0 = 1$; at first order we have $r_1 = \frac{1}{3}$ and $p_1 = \frac{8}{9}$, and at second order we have $r_2 = \frac{1}{9}$ and $p_2 = \frac{64}{81}$.

The probability that a square box of size r_i will include a retained square can be generalized to

$$p_i = N_i r_i^2 \tag{6.8}$$

and substitution of (2.1) gives

$$p_i = r_i^{2-D} \tag{6.9}$$

For the Sierpinski carpet the probability that a square box of size $r_i = (\frac{1}{3})^i$ will include a retained square is $p_i = (\frac{8}{9})^i$ so that $D = \ln 8/\ln 3$ as was previously found. The Sierpinski carpet can be applied to clustering in two dimensions in the same way that the Cantor set was applied in one dimension. This is directly analogous to the box-counting algorithm discussed in Chapter 2 and illustrated in Figure 2.8.

This approach can be extended to three dimensions using cubes of various sizes. The application to three dimensions is illustrated using the Menger sponge given in Figure 2.3(a). The objective is to determine the probability that a cube with size r encounters retained material. At zero order the probability that a cube of size $r_0 = 1$ will include material is $p_0 = 1$; at first order we have $r_1 = \frac{1}{3}$ and $p_1 = \frac{20}{27}$, and at second order we have $r_2 = \frac{1}{9}$ and $p_2 = \frac{400}{729}$. The probability that a cube of size r_i includes retained material can be generalized to

$$p_i = N_i r_i^3 \tag{6.10}$$

and substitution of (2.1) gives

$$p_i = r_i^{3-D} \tag{6.11}$$

For the Menger sponge the probability that a cube of size $r_i = (\frac{1}{3})^i$ encounters retained material is $p_i = \frac{20}{27}$ so that $D = \ln 20/\ln 3$ as was previously found. The generalization of (6.2), (6.9) and (6.11) is

$$p_i = r_i^{d-D} \tag{6.12}$$

where d is the Euclidean dimension of the problem being considered.

Problems

Problem 6.1. Consider the construction given in Figure 2.1(b). What is the probability that a step of length r includes a line segment for $r = 1, \frac{1}{3}, \frac{1}{9}, \frac{1}{27}$?

Problem 6.2. Consider the construction given in Figure 2.1(d). What is the probability that a step of length r includes a line segment for $r = 1, \frac{1}{3}, \frac{1}{9}, \frac{1}{27}$?

Problem 6.3. Consider the construction given in Figure 2.1(f). What is the probability that a step length r includes a line segment for $r = 1, \frac{1}{5}, \frac{1}{25}$?

Problem 6.4. A line segment is divided into seven equal parts and four are retained. The construction is repeated. What is the probability that a step of length r includes a line segment for $r = \frac{1}{7}, \frac{1}{49}, \frac{1}{343}$?

Problem 6.5. A line segment is divided into seven equal parts and three are retained. The construction is repeated. What is the probability that a step of length r includes a line segment for $r = \frac{1}{7}, \frac{1}{49}, \frac{1}{343}$?

Problem 6.6. Consider the construction given in Figure 2.2(c). What is the probability that a square box with dimensions r includes a retained square when $r = 1, \frac{1}{3}, \frac{1}{9}, \frac{1}{27}$?

Problem 6.7. A unit square is divided into four smaller squares of equal size. Two diagonally opposite squares are retained and the construction is repeated. What is the probability that a square box with dimensions r includes a retained square when $r = 1, \frac{1}{2}, \frac{1}{4}, \frac{1}{8}$?

Problem 6.8. A unit square is divided into 25 smaller squares of equal size. All the squares on the boundary and the central square are retained and the construction is repeated. What is the probability that a square box with dimensions r includes a retained square when $r = 1, \frac{1}{5}, \frac{1}{25}$?

Problem 6.9. Consider the construction given in Figure 2.3(b). What is the probability that a cube with dimensions r includes solid when $r = 1, \frac{1}{2}, \frac{1}{4}, \frac{1}{8}$?

Problem 6.10. A unit cube is divided into 27 smaller cubes of equal volume. All the cubes are retained except for the central one and the construction is repeated. What is the probability that a cube with dimensions r includes solid when $r = 1, \frac{1}{3}, \frac{1}{9}$?

CHAPTER SEVEN

Self-affine fractals

Up to this point we have considered self-similar fractals; we now turn to self-affine fractals (Mandelbrot, 1985). Topography is an example of both. In the two horizontal directions topography is often self-similar; the ruler method can be applied to a coastline or to a contour on a topographic map in order to define a fractal dimension. The box-counting method can also be applied to a coastline and square boxes are used to determine a fractal dimension. These are examples of self-similar fractals. Consider next the elevation of topography. Plots of elevation versus a horizontal coordinate resemble the random walk plots given in Figure 7.1. The vertical coordinate is statistically related to the horizontal coordinate but systematically has a smaller magnitude. Vertical cross-sections of this type are often examples of self-affine fractals (Dubuc *et al.*, 1989a).

A statistically self-similar fractal is by definition isotropic. In two dimensions defined by x and y coordinates the results do not depend on the geometrical orientation of the x- and y-axes. This principle was illustrated in Figure 2.8 where the box-counting method was

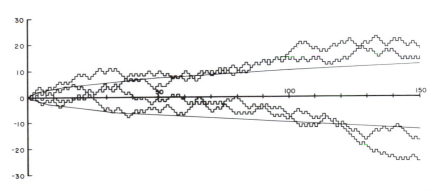

Figure 7.1. Four examples of a random walk (Brownian noise). After each step to the right a random vertical step is taken. One-hundred and fifty steps are illustrated. The continuous lines are the $n^{1/2}$ standard deviation for Brownian noise.

73

introduced. The fractal dimension of a rocky coastline is independent of the orientation of the boxes. A formal definition of a self-similar fractal in a two-dimensional xy-space is that $f(rx, ry)$ is statistically similar to $f(x, y)$ where r is a scaling factor. This result is quantified by applications of the fractal relation (2.6). The number of boxes with dimensions x_1, y_1 required to cover a rocky coastline is N_1; the number of boxes with dimensions $x_2 = rx_1$, $y_2 = ry_1$ required to cover a rocky coastline is N_2. If the rocky coastline is a self-similar fractal we have $N_2/N_1 = r^{-D}$.

A statistically self-affine fractal is not isotropic. A formal definition of a self-affine fractal in a two-dimensional xy-space is that $f(rx, r^H y)$ is statistically similar to $f(x, y)$ where H is known as the Hausdorff measure; we shall relate H to the fractal dimension in what follows. In the box-counting method square boxes become more and more rectangular as their size is increased.

An example of a self-affine fractal is given in Figure 7.1. This is an example of a random walk (Brownian noise). In order to generate a random walk take a step to the west and flip a coin; if heads you take a step to the left (south) and if tails you take a step to the right (north); take another step to the west and repeat the process. Four examples of a random walk are given in Figure 7.1. The box-counting methods can be used to determine the Hausdorff measure and fractal dimension of a random walk. However, the box sizes must be scaled using the Hausdorff measure. If N_1 is the number of boxes with dimensions x_1, y_1 required to cover the random walk and N_2 is the number of boxes with dimensions $x_2 = rx_1$, $y_2 = r^H y$, then the random walk is a self-affine fractal if $N_2/N_1 = r^{-D}$. We will show that the random walk is a self-affine fractal with $H = \frac{1}{2}$ and $D = \frac{3}{2}$. Before doing this, however, we will consider several other aspects of self-affine fractals.

Self-affine fractals are generally treated quantitatively using spectral techniques. We shall consider a single-valued function of time $x(t)$ that is random but has a specified spectrum. An example would be the value of the earth's magnetic field measured as a function of time at a point on the earth's surface. Such a time series is entirely equivalent to the dependence of topography on distance along a linear track. A fundamental consideration in time series is the correlation between the values $x(t + \tau)$ and $x(t)$. The larger the time interval τ the less the two values are expected to be correlated. In order for a time series to be a self-affine fractal it is necessary that the difference in the time series $x(t + \tau) - x(t)$ satisfies the probability condition

$$p\left[\frac{x(t + \tau) - x(t)}{\tau^H} < x'\right] = F(x') \qquad (7.1)$$

where the Hausdorff measure H is a constant. Point values of x are random but they are correlated with adjacent values by (7.1). If adjacent points are totally uncorrelated $H = 0$ and the result is white noise. The larger the value of H the smoother the dependence of x on t. In many examples $F(x')$ is a Gaussian distribution function, that is

$$F(x') = \frac{1}{2\pi}\int_0^{x'} \exp\left(-\frac{x^2}{2}\right) dx \qquad (7.2)$$

The values of x have a Gaussian distribution about $x = 0$ that is independent of the value H. If the distribution is Gaussian and if $0 < H < 1$ then the random signal is known as fractional Brownian noise; for $H = \frac{1}{2}$ we have Brownian noise.

The dependence of x on t in the random time series is similar to the shape of the Koch island illustrated in Figure 2.5. There are variations in the time series at all scales. Neither the length of the track defined by the time series nor the local derivatives (slopes) are defined. Thus it is appropriate to consider $x(t)$ to be a fractal. Just as in the case of the triadic Koch island and other fractal constructions on a plane the fractal dimension is between 1 and 2 and the appropriate Euclidian dimension is 2. With $d = 2$, (6.12) can be written

$$\frac{p}{\tau^{2-D}} = \text{constant} \qquad (7.3)$$

where the appropriate scale for the time series is τ. In (7.1) the time series diverges with the interval τ according to the power law τ^H. Comparing (7.1) and (7.3) we define

$$H = 2 - D \qquad (7.4)$$

This is the basic definition of the fractal dimension for a time series. Below we give an alternative derivation that gives the same result. For $1 < D < 2$ we require that $0 < H < 1$ which is the definition given above for fractional Brownian noise.

Assume that the time series $x(t)$ is specified over the interval of time T. It is useful to specify several statistical properties of the signal. The mean signal $\bar{x}(T)$ is given by

$$\bar{x}(T) = \frac{1}{T} \int_0^T x(t)\, dt \tag{7.5}$$

The variance of the signal $V(T)$ is defined by

$$V(T) = \frac{1}{T} \int_0^T [x(t) - \bar{x}]^2\, dt \tag{7.6}$$

The variance is directly related to the standard deviation of the signal $\sigma(T)$ by

$$\sigma(T) = [V(T)]^{1/2} \tag{7.7}$$

The mean and the variance are the first two moments of the time series.

A necessary condition that the time series be a fractal is that the variance $V(T)$ has a power-law dependence on T (Voss, 1985a, b, 1988):

$$V(T) \sim T^{2H} \tag{7.8}$$

or

$$\sigma(T) \sim T^H \tag{7.9}$$

This result is equivalent to (7.1) since T is equivalent to τ. The standard deviation of fractional Brownian noise increases as a fractional power of the interval of time T. For Brownian noise (random walk) $\sigma \sim T^{1/2}$ ($\sigma \sim n^{1/2}$) as illustrated in Figure 7.1.

An alternative derivation of the fractal dimension of a time series can be obtained by using the box-counting method. We first introduce a rectangular reference 'box' with a width T and height $\sigma_T = \sigma(T)$. Note that since the units of the signal x, and therefore the units of the standard deviation σ, can differ from the unit of time t, the aspect ratio (width/height) of the box can have arbitrary units. If we measure an electric current as a function of time the width of the box is in seconds and its height is in amperes.

We next divide the time interval T into n smaller time intervals with the length $T_n = T/n$. We also introduce scaled smaller boxes of width T_n, and height $\sigma_n = \sigma_T/n$. These boxes have the same aspect ratio as the reference box. However, the standard deviation associated with the interval T_n, $\sigma_{T_n} = \sigma(T_n) = \sigma(T/n)$, is not equal to σ_n. We determine the number of scaled smaller boxes N_n of size $T_n \times \sigma_n$ that are required to cover the area of width T and height σ_{T_n}. This is given by

$$N_n = \frac{T\sigma_{T_n}}{T_n\sigma_n} = n^2 \frac{\sigma_{T_n}}{\sigma_T} \tag{7.10}$$

Using (7.9) we have

$$\frac{\sigma_{T_n}}{\sigma_T} = \frac{\sigma(T/n)}{\sigma(T)} = \left(\frac{T/n}{T}\right)^H = \frac{1}{n^H} \tag{7.11}$$

and combining (7.10) and (7.11) gives

$$N_n = n^{2-H} = \left(\frac{T}{T_n}\right)^{2-H} \tag{7.12}$$

This is basically a fractal relation if we associate T_n with r_n. Comparing (7.12) with the definition of the fractal dimension in (2.1) we have the relation $2 - H = D$ which is identical to (7.4).

A time series can be either prescribed in the physical domain as $x(t)$ or in the frequency domain in terms of the amplitude $X(f, T)$ where f is the frequency. The quantity $X(f, T)$ is generally a complex number indicating the phase of the signal. The amplitude in the frequency domain, $X(f, T)$, is obtained using the Fourier transform of $x(t)$ in the interval $0 < t < T$; it is given by

$$X(f, T) = \int_0^T x(t) e^{2\pi i f t} \, dt \tag{7.13}$$

where $i = \sqrt{-1}$. The complementary equation relating $x(t)$ to $X(f, t)$ is the inverse Fourier transform

$$x(t) = \int_{-\infty}^{\infty} X(f, T) e^{-2\pi i f t} \, df \tag{7.14}$$

The quantity $|X(f, T)|^2 \, df$ is the contribution to the total energy of $x(t)$ from those components with frequencies between f and $f + df$. The vertical bars in $|X|$ refer to the absolute value of the complex quantity. The power is obtained by dividing by T. The power spectral density of $x(t)$ is defined by

$$S(f) = \frac{1}{T} |X(f, T)|^2 \tag{7.15}$$

in the limit $T \to \infty$. The product $S(f) \, df$ is the power in the time series associated with the frequency range between f and $f + df$.

For a time series that is a fractal the power spectral density has a power-law dependence on frequency:

$$S(f) \sim f^{-\beta} \tag{7.16}$$

We now obtain a relationship between the power β and the fractal dimension D.

We consider two time series $x_1(t)$ and $x_2(t)$ that are related by

$$x_2(t) = \frac{1}{r^H} x_1(rt) \tag{7.17}$$

The fundamental property of a self-affine fractal time series is that $x_1(t)$ has the same statistical properties as $x_2(t)$. The Fourier transform of $x_2(t)$ is given by

$$X_2(f, T) = \int_0^T x_2(t)e^{2\pi i f t}\, dt \tag{7.18}$$

Substituting (7.17) and making the change of variable $t' = rt$ we obtain

$$X_2(f, T) = \int_0^{rT} \frac{x_1(t')}{r^H} e^{2\pi i f t'/r} \frac{dt'}{r} \tag{7.19}$$

and comparing (7.19) with (7.13) we have

$$X_2(f, T) = \frac{1}{r^{H+1}} X_1\left(\frac{f}{r}, rT\right) \tag{7.20}$$

From the definition of the power spectral density given in (7.15) we obtain

$$S_2(f) = \frac{1}{T}|X_2(f, T)|^2 = \frac{1}{r^{2H+1}} \frac{1}{rT}\left|X_1\left(\frac{f}{r}, rT\right)\right|^2$$

$$= \frac{1}{r^{2H+1}} S_1\left(\frac{f}{r}\right) \tag{7.21}$$

Since x_2 is a properly rescaled version of x_1, their power spectral densities must also be properly scaled. Thus we can write

$$S(f) = \frac{1}{r^{2H+1}} S\left(\frac{f}{r}\right) \tag{7.22}$$

From (7.16) and (7.4) it follows that

$$\beta = 2H + 1 = 5 - 2D \tag{7.23}$$

For fractional Brownian noise $(0 < H < 1;\ 1 < D < 2)$ we have $1 < \beta < 3$. For Brownian noise $(H = \frac{1}{2},\ D = \frac{3}{2})$ we have $\beta = 2$.

We now consider some examples of fractional Brownian noise. These will be generated synthetically using the following steps:

(1) We consider $N = 500$ incremental steps of length ΔT. Each step is given a random value h_n based on the Gaussian probability distribution defined in (7.2). A typical example is illustrated in Figure 7.2(a). This is a Gaussian white noise sequence. Adjacent values are totally uncorrelated.

(2) A discrete Fourier transform is taken of the random values. The Fourier coefficients are given by

$$H_m = \Delta T \sum_{n=0}^{N-1} h_n e^{2\pi i n m/N} \tag{7.24}$$

This transform maps N real numbers (the h_n) into N complex numbers (the H_m). Because the transform is taken of a Gaussian white noise sequence the Fourier spectrum will be flat, that is $\beta = 0$ in (7.16). Except for the statistical scatter the amplitudes of the $|H_m|$ will be equal.

(3) The resulting Fourier coefficients H_m are filtered using the relation

$$H'_m = \left(\frac{m}{N-1}\right)^{\beta/2} H_m \qquad (7.25)$$

The power $\beta/2$ is used because the power spectral density is proportional to the amplitude squared. The amplitudes of the small-m coefficients correspond to short wavelengths λ_m and large wave numbers $k_m = 2\pi/\lambda_m$. The large-m coefficients correspond to long wavelengths and small wave numbers.

(4) An inverse discrete Fourier transform is taken of the filtered Fourier coefficients. The sequence of points is given by

$$h'_n = \frac{1}{N\,\Delta T} \sum_{m=0}^{N-1} H'_m \, e^{-2\pi i n m/N} \qquad (7.26)$$

These points constitute the fractional Brownian noise. In order to remove edge effects (periodicities) only the central portion should be retained.

Several examples of fractional Brownian noise are given in Figure 7.2. In each case the Gaussian white noise sequence given in Figure 7.2(a) has been filtered using the steps given above. As the value of β is increased the spatial correlation is increased and the profile is smoothed. Brownian noise ($\beta = 2$) is given in Figure 7.2(d); this result is statistically identical to the examples of random walk (Brownian motion) given in Figure 7.1.

There are many examples of time series. One example would be the measured temperature at a particular location. In order to better understand a time series it is standard procedure to carry out a spectral analysis. The spectral analysis of temperature would be expected to have strong spectral peaks at the frequencies associated with daily and annual periodicities. Such a time series would not be scale invariant.

Natural phenomena that yield continuous, power-law spectra do not have characteristic frequencies and are scale invariant over the applicable range. We will define these to be fractal. We will now consider several examples in geology and geophysics. One example is topography and bathymetry measured along a linear track

Self-affine fractals

Figure 7.2. Synthetically generated fractional Brownian noise. (a) Gaussian white noise sequence, (b) $\beta = 1.2$ ($D = 1.9$), (c) $\beta = 1.6$ ($D = 1.7$), (d) $\beta = 2.0$ ($D = 1.5$) (Brownian noise), (e) $\beta = 2.4$ ($D = 1.3$), (f) $\beta = 2.8$ ($D = 1.1$).

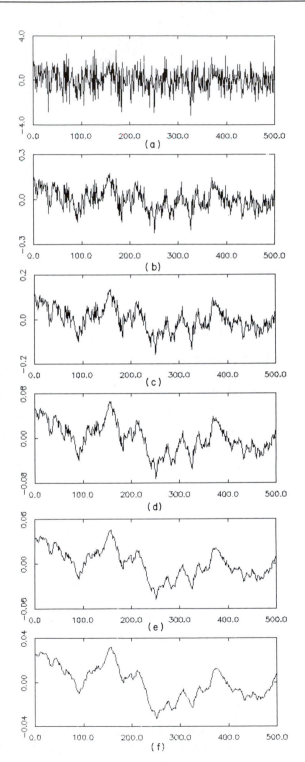

Table 7.1. *Regional averages over one-dimensional profiles of Oregon topography*

Area	Fractal dimension	Roughness
Willamette lowland		
Latitude	1.436	5.948
Longitude	1.507	6.354
Wallowa Mountains		
Latitute	1.499	6.549
Longitude	1.485	6.830
Klamath Falls		
Latitude	1.492	5.825
Longitude	1.500	5.963

(Bell, 1975, 1979; Berkson and Mathews, 1983; Barenblatt *et al.*, 1984; Fox and Hayes, 1985; Gilbert and Malinverno, 1988; Fox, 1989; Gilbert, 1989; Malinverno, 1989; Mareschal, 1989). Twenty-four examples from three different parts of Oregon are given in Figure 7.3. One-dimensional Fourier spectral analyses were obtained using the periodogram method. Three different regions were considered with different geomorphic and tectonic settings. The Willamette lowland is dominated by sedimentary processes, the Wallowa Mountains are associated with a major tectonic uplift, and the Klamath Falls area belongs to the basin and range tectonic regime. The topography was digitized along lines of latitude and longitude at seven points per kilometer. For each of the three regions, 20 equally spaced one-dimensional profiles of length 512 points were analysed in both the latitudinal and longitudinal directions. Log–log plots of the spectral power density versus wave number show a good power-law dependence in all three regions as shown in Figure 7.3. Eight typical examples are given for each of the three regions. The best-fit fractal dimension for each profile is obtained using (7.22). The mean fractal dimensions for three regions are given in Table 7.1. The mean values are close to $D = 1.5$ indicating that the spectral power density corresponds to Brownian noise to a good approximation. A variety of previous studies have found values for D near 1.5.

Two implications of this result will be discussed. The first is the

Figure 7.3. Plots of one-dimensional power spectral density versus wave number selected from three regions with different tectonic and geomorphic settings in Oregon. The profiles are offset vertically so as not to overlap. (a) Willamette lowland, (b) Wallowa Mountains, (c) Klamath Falls.

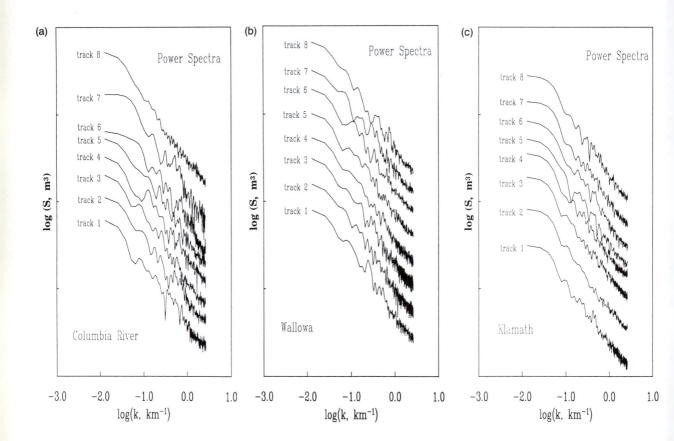

comparison with the value of *D* for topography obtained in Chapter 2 using the ruler method. As illustrated in Figure 2.7 the ruler method generally gives fractal dimensions near $D = 1.2$. These are systematically lower than the values near $D = 1.5$ obtained using the spectral method. Fundamentally there is no reason why the two fractal dimensions should be equal. Elevation profiles are not necessarily related to the shape of contours.

The correspondence of topography and bathymetry to Brownian noise also implies, importantly, that they are truly self-similar. For Brownian noise the amplitude coefficients are directly proportional to the corresponding wavelengths. Thus the height-to-width ratios of mountains and hills are the same at all scales.

It should also be noted that the power-law spectra given in Figure 7.3 provide further information beyond the fractal dimension. The spectra are characterized by the amplitude in addition to the slope. A quantitative measure of the amplitude is the intercept (value of *S*) at a specified wave number ($k = 1$ cycle km^{-1}). These reference amplitudes are a measure of the roughness of the topography. The mean intercepts for the latitudinal and longitudinal directions for the three regions in Oregon are given in Table 7.1.

Another application of spectral techniques is to well logs. It is common practice to make a variety of measurements as a function

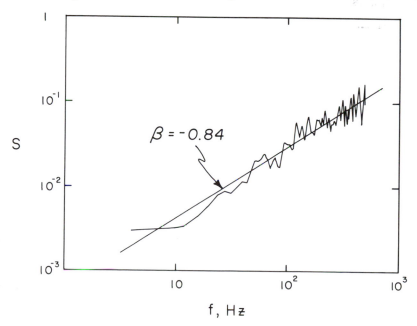

Figure 7.4. Power spectral density as a function of frequency for the reflection coefficients in the Amoco Puffin well on the Grand Banks of Newfoundland (Todoeshuck *et al.*, 1990). The straight line is the least-squares fit of (7.16) with $\beta = -0.84$.

of depth in oil wells. Typical measurements include the local acoustic velocity, the electrical conductivity, and neutron activation. The measured quantities are obtained as a function of depth so that they are equivalent to a time series and spectral techniques can be applied. An example is given in Figure 7.4. The power spectral densities of the reflection coefficients have been given for the Amoco Puffin well on the Grand Banks of Newfoundland (Todoeschuck *et al.*, 1990). These have been obtained from a sonic well log taken at depths of up to 3730 m and digitized at 1 ms intervals. The least-squares slope from (7.16) gives $\beta = -0.84$. The corresponding fractal dimension from (7.22) is $D = 2.92$. These values fall outside the range for fractional Brownian noise and fractal behavior. However, Walden and Hosken (1985) have pointed out that the power spectral density for the acoustic impedance is proportional to the power spectral density of the reflection coefficients divided by the square of the frequency. Thus the best-fit β for the acoustic impedance would be $\beta = 1.16$; the corresponding fractal dimension from (7.22) is $D = 1.92$. The acoustic impedance would be an example of fractional Brownian noise and fractal behavior. It appears that under many circumstances the variation of physical properties such as porosity in sedimentary sequences is fractal. The deposition of sedimentary sequences is dominated by major storms in which erosion occurs and previously deposited sediments are redeposited. One implication of fractal sequences is that the frequency–magnitude statistics of storms are also fractal. One measure of the severity of a storm is the magnitude of the resulting flood. In terms of flood control it is standard practice to specify the severity of the 100-year flood, the worst flood expected in 100 years. If fractal statistics are applicable then the ratio of the 1000-year flood to the 100-year flood is the same as the ratio of the 100-year flood to the 10-year flood and is the same as the ratio of the 10-year flood to the one-year flood. Fractal statistics generally predict more severe floods than exponential statistics, which are more commonly applied. Although much effort has gone into studying the frequency–magnitude statistics of floods, data are generally restricted to historical observations. It is extremely difficult to obtain accurate estimates of the magnitudes of paleo-floods.

There are many other examples of measurements in geology and geophysics that yield power-law spectra. Brown and Scholz (1985) have carried out spectral studies of natural rock surfaces. They generally find fractal behavior with a relatively large range of variability between $1 < D < 1.6$.

An interesting question is whether climate obeys fractal statistics (Nicolis and Nicolis, 1984). Fluigeman and Snow (1989) have shown that the spatial distribution of oxygen isotope ratios in sea floor cores obey fractal spectral statistics. Since it is generally accepted that the isotope ratios are proportional to the local temperatures, these results can be taken as evidence that climate obeys fractal statistics. Plotnick (1986) has argued that the distribution of stratigraphic hiatuses is fractal.

An important application of power-law spectra is in interpolating between measured data sets. Consider the bathymetry of the oceans. Bathymetry is typically measured from ships along linear tracks and must be interpolated to make bathymetric charts. This interpolation can make use of the fact that the bathymetry has a power-law spectrum. The amplitude coefficients are determined from the applicable fractal relation and the data are used to determine the phases in a two-dimensional Fourier expansion of the bathymetry. This method can also be used to interpolate airborne magnetic surveys.

Hewett (1986) has shown that porosity logs in oil fields obey fractal statistics. He has used three-dimensional fractal interpolations between well logs to determine the porosity structure of oil fields.

We now turn to two-dimensional spectral studies. Of particular interest is the earth's topography and bathymetry. Fractional Brownian noise can be visualized in terms of the earth's topography. Consider the elevation, h_2, of point 2 with respect to the elevation, h_1, of point 1. Point 1 has coordinates x_1 and y_1; point 2 has coordinates x_2 and y_2. For fractional Brownian noise we have

$$\Delta h \sim (\Delta r)^H \tag{7.27}$$

where $\Delta h = h_2 - h_1$, $\Delta r^2 = \Delta x^2 + \Delta y^2$, $\Delta x = x_2 - x_1$, and $\Delta y = y_2 - y_1$. Since topography is generally close to Brownian noise it is appropriate to take $H = \frac{1}{2}$ and the change of elevation is on average proportional to the square root of the distance walked, i.e. it is similar to Figure 7.1.

It is common practice to expand data sets on the surface of the earth in terms of spherical harmonics; examples include topography and geoid. Using topography as an example the appropriate expansion for the radius r of the earth is

$$r(\theta, \phi) = a_0 \left[1 + \sum_{l=1}^{\infty} \sum_{m=0}^{l} (C_{clm} \cos m\phi + S_{slm} \sin m\phi) P_{lm}(\sin \theta) \right] \quad (7.28)$$

where a_0 is a reference earth radius, θ is latitude, ϕ is longitude, C_{lm} and S_{lm} are coeffients, and P_{lm} are associated Legendre functions fully normalized so that

$$\frac{1}{4\pi} \int_0^{2\pi} \int_{-\pi/2}^{\pi/2} P_{lm}(\sin \theta) \begin{Bmatrix} \cos^2 m\phi \\ \sin^2 m\theta \end{Bmatrix} \mathrm{d}(\sin \theta) \, \mathrm{d}\phi = 1 \quad (7.29)$$

The variance of the spectra for order l is defined by

$$V_l = a_0^2 \sum_{m=0}^{l} (C_{clm}^2 + C_{slm}^2) \quad (7.30)$$

and the power spectral density is defined by

$$S(k) = \frac{1}{k_0} V(k) = \lambda_0 V(k) \quad (7.31)$$

where k_0 is the wave number and $\lambda_0 = 1/k_0$ is the wavelength over which data is included in the expansion. With $\lambda_0 = 2\pi a_0$ we have

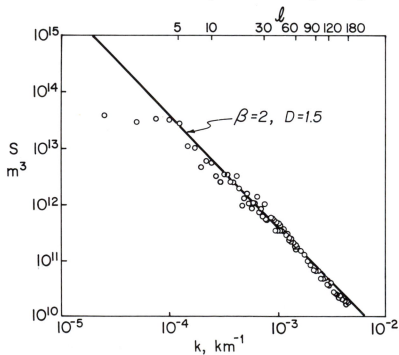

Figure 7.5. Power spectral density of the earth's topography as a function of wave number. The circles are from the data compilation of Rapp (1989). The solid line represents (7.16) with $\beta = 2$ ($D = 1.5$).

$$S_l = 2\pi a_0^3 \sum_{m=0}^{l} (C_{clm}^2 + C_{slm}^2) \qquad (7.32)$$

$$k_l = \frac{l}{2\pi a_0} \qquad (7.33)$$

A fractal dependence can be defined if S_l has a power-law relation to the wave number k_l.

The power spectral density of the earth's topography as a function of wave number is given in Figure 7.5. This is based on the spherical harmonic expansion of the earth's topography to order $l = 180$ by Rapp (1989). Except for the low-degree harmonics an excellent correlation is obtained with (7.16) taking $\beta = 2 (D = 1.5)$. The spectral dependence of topography corresponds to Brownian noise as previously noted. The power spectral density of the earth's geoid is given as a function of wave number in Figure 7.6; this was compiled by Turcotte (1987) from the data given by Reigber *et al.* (1985). Except for the

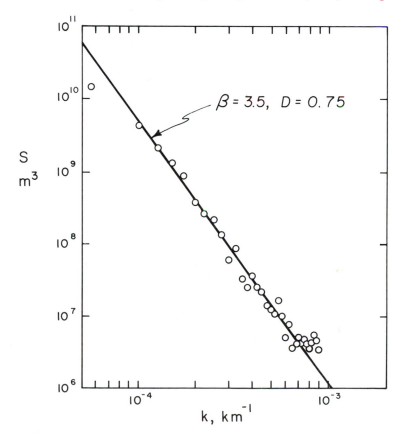

Figure 7.6. Power spectral density of the earth's geoid as a function of wave number. The circles represent a compilation of the data of Reigber (1977). The solid line represents (7.16) with $\beta = 3.5$ $(D = 0.75)$.

low-degree harmonics an excellent correlation is obtained with (7.16) taking $\beta = 3.5$ ($D = 0.75$). This is another example that falls outside the range of fractional Brownian noise and the formal limits for fractal behavior.

The Fourier spectral approach to fractal analysis for one-dimensional profiles discussed previously can be extended to two-dimensional image analysis (Dubuc *et al.*, 1989b). Consider an $N \times N$ grid of equally spaced data points in a square with linear size L. The N^2 data points are denoted by h_{nm} with (n, m) specifying the position in the x- and y-directions respectively. A case with $N = 8$ is illustrated in Figure 7.7(a).

The first step is to carry out a two-dimensional discrete Fourier transform on the N^2 set of data points h_{nm}. An $N \times N$ array of complex coefficients H_{st} is obtained by the usual definition

$$H_{st} = \left(\frac{L}{N}\right)^2 \sum_{n=0}^{N-1} \sum_{m=0}^{N-1} h_{nm} \exp\left[-\frac{2\pi i}{N}(sn + tm)\right] \qquad (7.34)$$

where s denotes the transform in the x-direction ($s = 0, 1, 2, \ldots, N - 1$) and t denotes the transform in the y-direction ($t = 0, 1, 2, \ldots, N - 1$). Then each transform coefficient H_{st} is assigned an equivalent radial number using the relation

$$r = (s^2 + t^2)^{1/2} \qquad (7.35)$$

The two-dimensional mean power spectral density S_{2j} for each radial wave number k_j is given by

$$S_{2j} = \frac{1}{L^2 N_j} \sum_{1}^{N_j} |H_{sl}|^2 \qquad (7.36)$$

where N_j is the number of coefficients that satisfy the condition $j < r < j + 1$ and the summation is carried out over the coefficients H_{st} in this range. The coefficients assigned to each interval for the example given in Figure 7.7(a) are illustrated in Figure 7.7(b).

The dependence of the mean power spectral density on the radial wave number k_j for a fractal distribution is (Voss, 1988)

$$S_{2j} \sim k_j^{-\beta - 1} \qquad (7.37)$$

instead of (7.16). The addition of minus one to the power k_j is required because of the radial coordinates that are used in phase space. The dependence of $V(L)$ on L given in (7.8) is still valid but with the additional dimension the 'box' derivation that follows now gives

$$H = 3 - D_2 \tag{7.38}$$

for the fractal dimension of the surface instead of (7.4). Similarly, the derivation of the relationship between β and H must be re-examined but

$$\beta = 2H + 1 \tag{7.39}$$

	1 8	2 8	3 8	4 8	5 8	6 8	7 8	8 8
	1 7	2 7	3 7	4 7	5 7	6 7	7 7	8 7
	1 6	2 6	3 6	4 6	5 6	6 6	7 6	8 6
m	1 5	2 5	3 5	4 5	5 5	6 5	7 5	8 5
	1 4	2 4	3 4	4 4	5 4	6 4	7 4	8 4
	1 3	2 3	3 3	4 3	5 3	6 3	7 3	8 3
	1 2	2 2	3 2	4 2	5 2	6 2	7 2	8 2
	1 1	2 1	3 1	4 1	5 1	6 1	7 1	8 1

n

(a) The 64 *nm* coefficients for an 8 x 8 sub-set of raw data.

	8	8	8	8	8				
	7	7	7	7	8	8			
	6	6	6	7	7	7	8		
	5	5	5	5	6	7	7	8	
t	4	4	4	5	5	6	7	8	8
	3	3	3	4	5	5	7	7	8
	2	2	2	3	4	5	6	7	8
	1	1	2	3	4	5	6	7	8
	0	1	2	3	4	5	6	7	8

s

(b) Equivalent radial coefficients *r* for various coefficients *s* and *t* in spatial frequency space.

Figure 7.7. Illustration of subscript arrangement in two-dimensional spectral analysis.

remains valid. Combining (7.38) and (7.39) gives

$$D_2 = \frac{7 - \beta}{2} \qquad\qquad (7.40)$$

for the two-dimensional case.

Synthetic images can be generated using the same technique used to generate synthetic fractional Brownian noise. The method used to generate images is as follows.

(1) We consider an $N \times N$ square grid consisting of N^2 equally spaced points. Each point is given a random value h_{mn} based on the Gaussian probability distribution defined in (7.2). A typical example is illustrated in Figure 7.8(a). This is Gaussian white noise so that adjacent points are totally uncorrelated and $\beta = 0$.

(2) A two-dimensional discrete Fourier transform is taken using (7.34) generating an $N \times N$ array of complex coefficients H_{st}.

(3) A fractal dimension D_2 is specified and the corresponding value for β is obtained from (7.40). A new set of complex coefficients are obtained from the relation

$$H_{st}^* = H_{st}/k_r^{\beta/2} \qquad\qquad (7.41)$$

(4) An inverse two-dimensional discrete Fourier transform is carried out to generate a new image.

Five examples of synthetically generated images are given in Figure 7.8. In (b) $\beta = 1.2$ ($D_2 = 2.9$), in (c) $\beta = 1.6$ ($D_2 = 2.7$), in (d) $\beta = 2.0$ ($D_2 = 2.5$), in (e) $\beta = 2.4$ ($D_2 = 2.3$), and in (f) $\beta = 2.8$ ($D_2 = 2.1$). The synthetic result for $D_2 = 2.5$ looks quite realistic for a typical topographic map.

We have carried out one-dimensional spectral decompositions of linear profiles of our synthetic data. The results for synthetic topography with $D_2 = 2.6$, 2.7 and 3.0 are given in Table 7.2. For realistic topography with $D_2 = 2.6$ we find that the corresponding one-dimensional profiles give $D_2 = 1.58$. This is consistent with the previously published results for one-dimensional bathymetric and topographic profiles where values near $D = 1.5$ have been found as discussed above. In general the relation $D_2 = D + 1$ is a good approximation.

Table 7.2. *Summary of mean fractal dimensions estimated by one-dimensional and two-dimensional spectral analysis for the topography of Oregon and for synthetic images*

Data	Average D	
	Two-dimensional analysis	One-dimensional analysis
Oregon topography	2.586	1.487
Synthetic topography	2.60	1.58
	2.70	1.65
	3.00	1.91

The results given above can be extended in order to carry out fractal mapping of digitized images. As a specific example we consider the digitized topography of the state of Oregon (Huang and Turcotte, 1990a). Combining Defense Mapping Agency (DMA) 1° × 1° data with topographic maps, the US Geological Survey (Flagstaff) has produced digitized topography on a grid scale of about seven points per kilometer. The topography for Oregon is illustrated in Figure 7.9.

The fact that fractal statistics are a good approximation for topography allows us to make fractal maps of a region of diverse tectonics. Using the digitized topography of Oregon, plots of power spectral density versus wave number are made for subregions. From these plots a fractal dimension (slope) and unit wave number amplitude are obtained for each subregion. The amplitude is a measure of roughness. We are basically carrying out a texture analysis using the fractal statistics as a basis.

Fractal dimensions and roughness amplitudes are obtained using subregions of 32 × 32 data points. Thus fractal dimensions and roughness amplitudes are obtained for each 4.5 × 4.5 km subregion in the state; maps are generated. The 32 × 32 set was chosen because it generally gives well defined fractal spectra; for smaller regions the errors in fractal dimension and roughness becomes substantially larger. For larger regions, the spatial resolution of the map is degraded.

The following technique is used to obtain a fractal dimension and roughness amplitude for each subregion.

(1) A 32×32 set of digitized elevations is chosen to form each subregion ($N = 32$).
(2) The mean and linear trends for each subset of data are removed.
(3) A two-dimensional discrete Fourier transform is carried out, and an $N \times N$ array of complex Fourier coefficients H_{st} is obtained using (7.34).
(4) Each coefficient H_{st} is assigned an equivalent radial wave number r using (7.35). The two-dimensional mean spectra energy density S_{2j} is obtained for each radial integer wave number k_j using (7.36).
(5) The mean slope on a log–log plot of S_{2j} versus k_j obtained by a least-squares regression yields a fractal dimension D_2 using (7.37) and (7.40); the intercept at $k_j = 1$ cycle km^{-1} yields a roughness amplitude.

Examples for four randomly selected subregions in Oregon are given in Figure 7.10. The mean two-dimensional fractal dimension for all of Oregon is $D_2 = 2.586$ (Table 7.2). This is remarkably close to the mean value $D_2 = 2.59$ that was obtained for the state of Arizona (Huang and Turcotte, 1989). It is seen that the results are in good agreement with the relation $D_2 = 1 + D$.

Maps of fractal dimension and roughness amplitude are given in

Figure 7.10. Plots of mean power spectral density versus radial wave number for four typical 32×32 point subregions in Oregon. The linear trend on the log–log plot indicates a power-law (fractal) distribution.

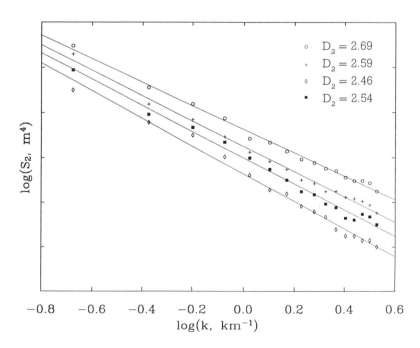

Figure 7.11. As expected, there is relatively little variation in the fractal dimension about the mean value, although the range is from about $2.40 < D < 2.90$. The variation in the roughness amplitude in Figure 7.11(b) is much more impressive. The sedimentary Willamette lowland shows low overall roughness while the erosion system associated with the nearby mountain ranges and the Wallowa Mountains in the northeast stand out as regions of high roughness. The roughness contrasts in the southern basin and range region are also quite remarkable. The fractal analysis gives a quantitative measure of roughness.

Problems

Problem 7.1. The definition of red noise is $\beta = 1$. What is the fractal dimension? How do the variance V and standard deviation σ depend upon the interval T? Is red noise an example of fractional Brownian noise; if so, why?

Problem 7.2. Determine the aspect ratio (height-to-width ratio of the mountains and valleys) using the correlation line from Figure 7.4. For this case the spectral power density is defined as the amplitude coefficient squared times the circumference of the earth.

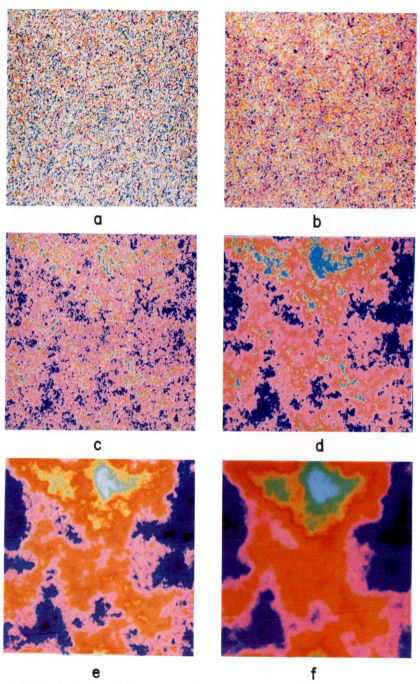

Figure 7.8. Synthetic fractal images on a 256 × 256 grid. (a) White noise without fractal filtering. (b) Filtered white noise with ß = 1.2 and D = 2.9. (c) Filtered white noise with ß = 1.6 and D = 2.7. (d) Filtered with ß = 2.0 and D = 2.5. (e) Filtered with ß = 2.4 and D = 2.3. (f) Filtered with ß = 2.8 and D = 2.1.

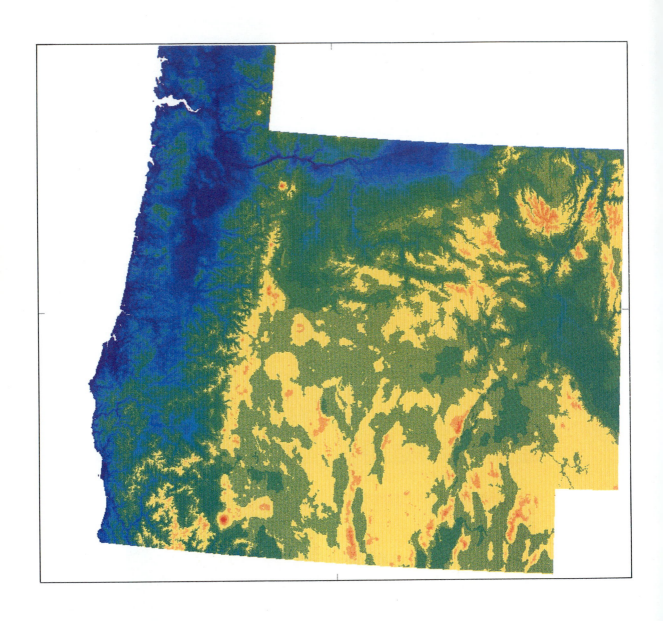

Figure 7.9. Map of the digitized topography for Oregon. Data resolution is about seven points per kilometer. The width of the state of Oregon is about 375 miles.

(a)

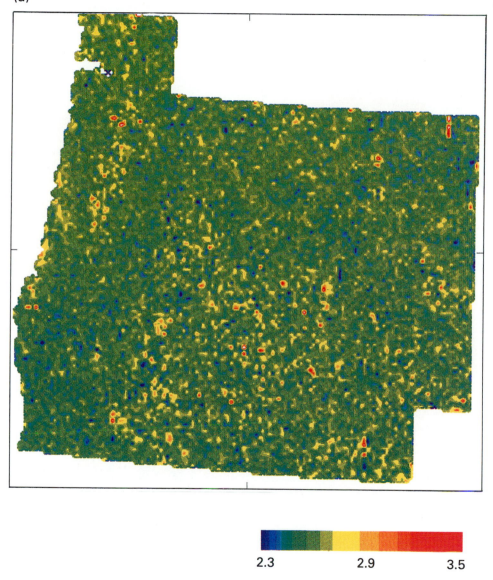

2.3   2.9   3.5

Figure 7.11. Maps of (a) fractal dimension and (b) roughness amplitude for Oregon. There is generally limited systematic variation in the fractal dimension; however, the roughness amplitude is sensitive to texture changes.

(b)

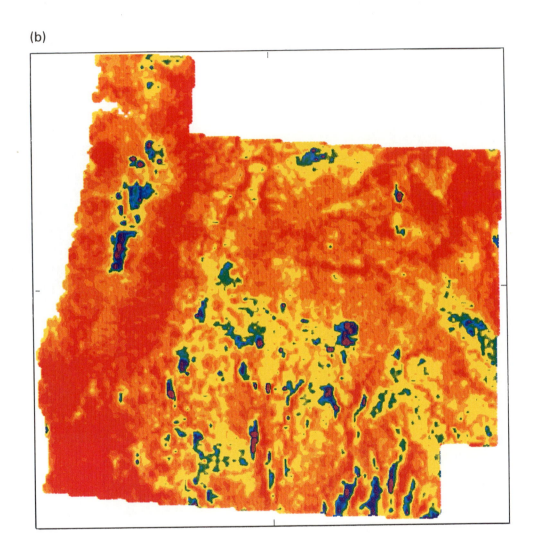

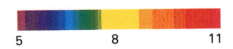

5   8   11

CHAPTER EIGHT

Geomorphology

In the previous chapters we concluded that landscapes generally obey fractal statistics. This is evidence that the mechanisms responsible for landscape evolution are scale invariant but it does not constrain the mechanism. The primary mechanism responsible for the evolution of topography is erosion. There are several aspects to erosion. The first is the development of soils and rock fragments through a combination of chemical and mechanical weathering. One example is freezing and thawing. The soils and rock fragments are then transported as sediments in rivers and streams. The rivers and streams themselves erode channels and gullies to form drainage patterns.

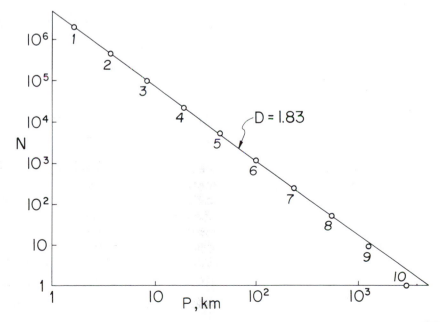

Figure 8.1. Dependence of the number N of rivers with a specified order on their mean length P. The correlation with (2.1) gives $D = 1.83$.

Under many circumstances drainage patterns also yield fractal statistics. In Figure 8.1 the number of rivers in the United States of a given order is plotted against the mean length of those rivers. A first-order river has no tributaries on a standard topographic map, a second-order river has at least one tributary (branch), a third-order river has at least one tributary that is second order, and so forth. It is seen that the number–order statistics given in Figure 8.1 correlate well with the fractal relation (2.1) taking $D = 1.83$.

Another fractal correlation to drainage patterns is obtained if the length of the principal river in a drainage basin, P, is plotted against the area of the basin, A. Data for several basins in the northeastern United States are given in Figure 8.2 (Hack, 1957). The applicable fractal relation is

$$P = CA^{D/2} \tag{8.1}$$

and good agreement with the data in Figure 8.2 is obtained taking $D = 1.22$. In young terrains tectonic processes play an important role; however, erosional processes may still be dominant. Consider the Hawaiian chain of volcanic islands. A young island such as Hawaii is made up of deterministic conical structures associated with shield volcanoes. These are not fractal. However, sufficient erosion has occurred on Maui and Oahu in a few million years to develop an irregular, scale-invariant morphology that exhibits fractal statistics. The erosional evolution of landscapes is a problem that has fascinated natural scientists for centuries. The forms of mature landscapes evolve

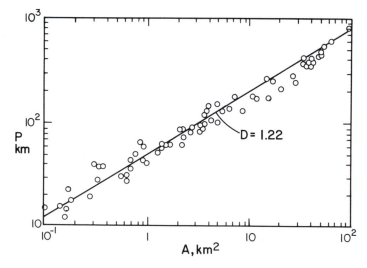

Figure 8.2. Dependence of the length P of the principle river on the area A of the drainage basin for several drainage basins in the northeastern United States (Hack, 1957).

through processes of erosion and deposition. An essential question is whether it is possible to develop a basic theory of landscapes or whether it is necessary to consider only statistical aspects of the problem.

A number of theories have been proposed for erosion and have been summarized by Scheidegger (1970). The simplest theory hypothesizes that erosion is proportional to the elevation h so that

$$\frac{\partial h}{\partial t} = -\frac{h}{\tau} \tag{8.2}$$

where τ is a characteristic time for erosion. With the initial condition that $h = h_0$ at $t = 0$ the solution is

$$h = h_0 \exp(-t/\tau) \tag{8.3}$$

This gives an exponential decay of topography with a characteristic time τ but no information on the form of the evolving landscapes.

Culling (1960, 1963, 1965) hypothesized that the horizontal flux of eroded material $\dot{m}_x$ is proportional to the slope:

$$\dot{m}_x = -\rho J \frac{\partial h}{\partial x} \tag{8.4}$$

where J is the transport coefficient and we consider only the one-dimensional problem $h = h(x, t)$. With the conservation of mass relation

$$\rho \frac{\partial h}{\partial t} + \frac{\partial \dot{m}_x}{\partial x} = 0 \tag{8.5}$$

this gives

$$\frac{\partial h}{\partial t} = J \frac{\partial^2 h}{\partial x^2} \tag{8.6}$$

which is the linear diffusion equation.

Solutions to the diffusion equation take a variety of forms but in many cases reduce to the error function. The form of many alluvial fans and prograding deltas can be approximated quite accurately with the error function (Kenyon and Turcotte, 1985). In addition, a number of authors have used the Culling model to estimate the age of faults and shoreline scarps (Wallace, 1977; Buckman and Anderson, 1979; Nash, 1980a, b; Mayer, 1984; Hanks *et al.*, 1984; Hanks and Wallace, 1985; Andrews and Hanks, 1985). Typical values for J are in the range 10^{-2}–$10^{-3}\,\mathrm{m^2\,yr^{-1}}$.

No linear theory will produce self-similar topography. As an example, we consider the Culling model given in (8.6). We assume that topography has a Fourier representation

$$h(x,t) = \int_{-\infty}^{\infty} H(k,t)e^{-2\pi ikx}\,dk \qquad (8.7)$$

where k is the wave number. Substitution into (8.6) yields

$$H(k,t) = H(k,0)e^{-4\pi^2 Jk^2 t} \qquad (8.8)$$

The amplitudes decay exponentially and the rate of decay is proportional to the square of the wave number. Short-wavelength topography (large k) decays much more rapidly than long-wavelength topography (small k). The power spectral density of the topography is proportional to the square of the spectral amplitude so that

$$\frac{S(k,\tau)}{S(k,0)} = \exp\left(-\frac{2\tau k^2}{k_0^2}\right) \qquad (8.9)$$

where $\tau = 4\pi^2 Jk_0^2 t$ and k_0 is a reference wave number. We assume the power spectral density of the initial topography is inversely proportional to the square of the wave number (Brownian noise) so that

$$\frac{S(k,0)}{S(k_0,0)} = \left(\frac{k_0}{k}\right)^2 \qquad (8.10)$$

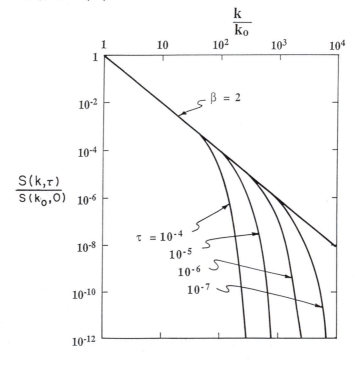

Figure 8.3. Power spectral density as a function of wave number for initial power-law topography (8.10) eroded according to the Culling model (8.11) for four erosion times τ (8.10). The short-wavelength topography (large k) is rapidly eroded.

Combining (8.9) and (8.10) gives

$$\frac{S(k,\tau)}{S(k_0,0)} = \left(\frac{k_0}{k}\right)^2 \exp\left(-\frac{2\tau k^2}{k_0^2}\right) \qquad (8.11)$$

The dependence of $S(k,\tau)/S(k_0,0)$ on k/k_0 for several values of τ is given in Figure 8.3. The strong decay of topography with large wave numbers (small wavelengths) is clearly illustrated; it destroys the initial self-similarity (fractal behavior). The linear diffusion equation (8.6) is not self-similar; it includes a characteristic length scale $(Jt)^{1/2}$, and therefore the evolution of an initial distribution is not self-similar (is not fractal). The decay shown in Figure 8.3 is not typical of actual topography; see for example Figure 7.3.

It is reasonable to conclude that the Culling (diffusion) equation does not generally govern the evolution of landscapes. An alternative approach that preserves self-similar topography will now be discussed. In order to avoid mathematical difficulties we will consider topography relative to its mean height $\langle h(x,t)\rangle$ where the brackets $\langle$ and $\rangle$ denote averages over many samples of $h(x,t)$. The autocovariance function for this topography is defined by

$$\rho(X,t) = \langle h(x+X,t)h(x,t)\rangle \qquad (8.12)$$

which is presumed to be independent of x (i.e. it is spatially homogeneous). The Fourier representation for $\rho(X,t)$ relates the autocovariance function to the power spectral density according to

$$\rho(X,t) = \int_{-\infty}^{\infty} S(k,t)\,e^{-2\pi i k X}\,dk \qquad (8.13)$$

In order to model a geomorphology cascade, Newman and Turcotte (1990) introduced a modified Fourier series associated with topography $h(x,t)$ on a linear track

$$\rho(X,t) = \sum_{n=0}^{\infty} S_n(t)\exp(-2\pi i 2^n k_0 X) \qquad (8.14)$$

with $N = 0,1,2,\ldots$. Unlike conventional Fourier series representations, this 'renormalized' representation compresses all the information contained between wave numbers $k_0 2^{n-1/2}$ and $k_0 2^{n+1/2}$ into the 'Fourier amplitude coefficient' S_n. This series representation, although not invertible or complete, is scale invariant in the sense that when we compare a coefficient, say S_n, with an adjacent one, say S_{n+1}, we are comparing two different scales and the index n can be regarded as being a logarithmic representation of scale. This procedure is conceptually the same as that employed by Kolmogorov (1941) in the theory of fully developed isotropic turbulence.

It is of interest to relate the coefficients S_n in the modified Fourier series (8.13) to the spectrum $S(k)$ in the standard Fourier integral representation (8.12). In order to do this we associate the spectrum $S(k)$ in the region $k_0 2^{n-1/2} < k < k_0 2^{n+1/2}$ with the coefficient S_n so that

$$S_n(t) = \int_{k_0 2^{n+1/2}}^{k_0 2^{n-1/2}} S(k, t)\, dk \tag{8.15}$$

For convenience, we now define

$$k_n = k_0 2^n \tag{8.16}$$

Considering power law spectra that satisfy (7.16), the expression for $S(k, t)$ is

$$S(k, t) = S(k_0, t)\left(\frac{k}{k_0}\right)^{-\beta} \tag{8.17}$$

substitution of (8.17) into (8.15) and integration gives

$$S_n(t) = \frac{S(k_0, t)}{(1-\beta)} k_0^{\beta} (2^{-(1-\beta)/2} - 2^{(1-\beta)/2}) k_n^{1-\beta} \tag{8.18}$$

We observe that our scaled coefficients satisfy

$$\frac{S_{n+1}}{S_n} = 2^{1-\beta} \equiv \lambda \tag{8.19}$$

since

$$\frac{k_{n+1}}{k_n} = 2 \tag{8.20}$$

Using the relation between β and D given in (7.23) we obtain

$$\lambda \equiv 2^{2D-4} = 2^{2(D-2)} \tag{8.21}$$

The ratio given in (8.19) can also be directly related to fractional Brownian noise. The coefficient S_n is a measure of the mean-squared height of features with wave numbers in the vicinity of k_n. We will assume for the moment that Δr in (7.27) corresponds to the wavelength $\lambda_n \approx 2\pi k_n^{-1}$. From (7.27) and (7.4) we obtain

$$\frac{S_{n+1}}{S_n} = \left[\frac{\Delta h(\Delta r/2)}{\Delta h(\Delta r)}\right]^2 = \left[\frac{\Delta r}{2\Delta r}\right]^{2H} = \left(\frac{1}{2}\right)^{2H} = 2^{2(D-2)} \tag{8.22}$$

which is in agreement with (8.19) and (8.21). When the parameter $H = 0$, the fractal dimension $D = 2$, the spectral coefficients S_n are equal, and $\beta = 1$ so that the spectrum $S(k)$ describes $(1/f)$-noise. For

Brownian noise $\beta = 2$, $D = 1.5$, $H = \frac{1}{2}$, and $\lambda = \frac{1}{2}$.

In order to study how fractal topography may evolve we hypothesize that the Fourier coefficients S_n satisfy the equation

$$\frac{dS_n}{dt} = -R_n S_n \tag{8.23}$$

where R_n is the rate coefficient appropriate to that scale. Since we restrict our attention to erosional processes with no active tectonics, it is appropriate to assume that R_n is positive. The effect is that spectral coefficients in the absence of sources must decay. The simple model (8.2) is a special case of this equation. In particular, it corresponds to (8.23) if $R_n = 2/\tau$ for all n. This feature illustrates the principal ingredient of the renormalization method that is not available to the linear model (8.2): erosion rates at every length scale depend upon the topographic distribution of features at related length scales.

In many nonlinear problems it is appropriate to assume that the rate coefficient R_n is a function of the Fourier coefficients at that scale, S_n, and at the immediately adjacent scales, S_{n+1} and S_{n-1}. This is equivalent to the Kolmogorov–Obukhov turbulent cascade scaling relation in the inertial subrange (Landau and Lifshitz, 1959).

For the erosion problem, we will specify relations for the R_n based on the observation that large gullies generate gullies on the next smaller scale and so forth in a scale-invariant way. At the largest scale, $n = 0$ and $k = k_0$, we take $R_0 = \alpha$, a constant; thus

$$\frac{dS_0}{dt} = -\alpha S_0 \tag{8.24}$$

Consequently, the largest features have an exponential decay as is often assumed in the literature. At other scales, we assume that features of a given scale are only influenced by the next larger scale. This would correspond to a gully of a specified scale causing smaller (by a factor of about two) gullies to emerge by erosion, the process being scale-invariant. Thus we assume that R_n is a function only of S_{n-1} and S_n. It is reasonable to assume that R_n approaches a constant value γ as S_n/S_{n-1} approaches a constant value λ. We thus consider the model relation

$$R_n = \gamma \left(\frac{S_n}{\lambda S_{n-1}} \right) \tag{8.25}$$

where γ and λ are constants. The bracketed term could be raised to an arbitrary power so that an infinite set of models could be considered. (It can be shown, however, that the qualitative features of the results that we shall derive shortly are exactly preserved.) It is easy to verify that if the amplitudes S_n are initially self-similar, that is if

$$S_n(0) = \lambda S_{n-1}(0) \tag{8.26}$$

for $n = 1, 2, 3, \ldots$, then the rate coefficients R_n for $1, 2, 3, \ldots$ are the same (i.e. independent of n), specifically

$$R_n = \gamma \tag{8.27}$$

However, the rate coefficient for the largest scale is $R_0 = \alpha$; therefore, for this model to remain consistent we require that

$$\alpha = \gamma \tag{8.28}$$

Thus, in scale-invariant equilibrium, the rate coefficients at all scales are equal.

We have not provided a detailed physical basis for our model. We argue that it represents the creation of erosional features in major storms. We believe that all erosional landscapes are characterized by drainage features that are continuously being renewed. This renewal takes place in the largest storms, which generate the largest floods. An extreme example would be glaciation.

An interesting separate, but related, question is whether storms and in particular floods have a fractal distribution. If floods have a fractal distribution the ratio of the greatest mean flood in 1000 years to the greatest mean flood in 100 years would be equal to the ratio of the 100-year flood to the 10-year flood, etc. An associated question is whether a large fraction of erosion takes place in the very largest floods. We would argue that the fractal structure is evidence for a fractal distribution of erosion, i.e. of great floods.

Problem

Problem 8.1. Consider the fractal model for a river network illustrated in
Figure 8.4. Assume scale invariance such that $h_1/h_2 = h_2/h_3 = R$.
(i) Show the length P_n of the nth-order river is given by

$$P_n = h_1 + h_2 + h_3 + h_4 + \cdots = \frac{Rh_1}{R-1}$$

and that the length P_{n-1} of the $(n-1)$th-order river is given by

$$P_{n-1} = h_2 + h_3 + h_4 + \cdots = \frac{Rh_2}{R-1}$$

(ii) Since $N_n = 1$, $N_{n-1} = 2$, $N_{n-2} = 4$, etc. show that $D = \ln 2/\ln R$.
Determine R for the example given in Figure 8.4 ($h_2 = h_1/2$, etc.).

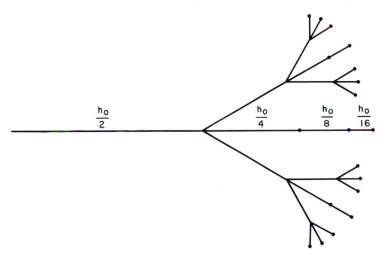

Figure 8.4. Model for a
fractal river network.

CHAPTER NINE

Dynamical systems

We now turn our attention to some examples of deterministic chaos that have applications in geology and geophysics. There are two requirements for a solution that exhibits deterministic chaos. The first requirement is that we are solving deterministic equations with specified initial and/or boundary conditions. Thus the applicable equations are deterministic not statistical. The second requirement is that solutions that have initial conditions that are infinitesimally close diverge exponentially as they evolve. However, before we consider solutions that are chaotic, some necessary introductory material is presented. Chaotic behavior is only found for nonlinear systems of equations. In this chapter some of the standard nomenclature for the study of nonlinear equations is presented (Verhulst, 1990).

Probably the simplest nonlinear total differential equation is the logistic equation

$$\frac{dx}{dt} = \frac{x}{\tau}\left(1 - \frac{x}{x_e}\right) \tag{9.1}$$

This equation is a simple model for population dynamics with x representing population and t time; the first term on the right-hand side accounts for births and the second deaths. The parameter τ is the characteristic time and x_e is a representative population. It is standard procedure in treating nonlinear equations to introduce nondimensional variables. This procedure reduces the number of parameters that must be varied in order to study the solutions. For the logistic equation it is appropriate to introduce

$$\bar{x} = \frac{x}{x_e}, \quad \bar{t} = \frac{t}{\tau} \tag{9.2}$$

Substitution of (9.2) into (9.1) gives

$$\frac{d\bar{x}}{dt} = \bar{x}(1 - \bar{x}) \tag{9.3}$$

and there are no governing parameters. An exact solution of (9.3) is

$$\bar{x} = \frac{\bar{x}_0 e^{\bar{t}}}{1 - \bar{x}_0(1 - e^{\bar{t}})} \tag{9.4}$$

where $\bar{x} = \bar{x}_0$ at $\bar{t} = 0$ is the initial condition. Solutions of (9.4) for various values of $\bar{x}_0$ are given in Figure 9.1.

Although (9.1) is solved by writing (9.4) it is illustrative to examine the solutions in somewhat more detail. We first examine the fixed points $\bar{x}_p$ of (9.3). As the solutions of (9.3) evolve in time the time dependence dies out and $\bar{x}$ approaches a particular value. This value is known as a stable fixed point. The fixed points are formally obtained by setting $d\bar{x}/d\bar{t} = 0$ in (9.3) with the result that

$$\bar{x}_p = \begin{cases} 0 \\ 1 \end{cases} \tag{9.5}$$

If the initial condition is either of these values then no time dependence is obtained. It is also of interest to examine the stability of the fixed points. In order to do this the applicable equation is linearized about the fixed point. In the vicinity of the fixed point $\bar{x}_p = 0$ it is appropriate to neglect the quadratic term in (9.3) with the result

$$\frac{d\bar{x}}{d\bar{t}} = \bar{x} \tag{9.6}$$

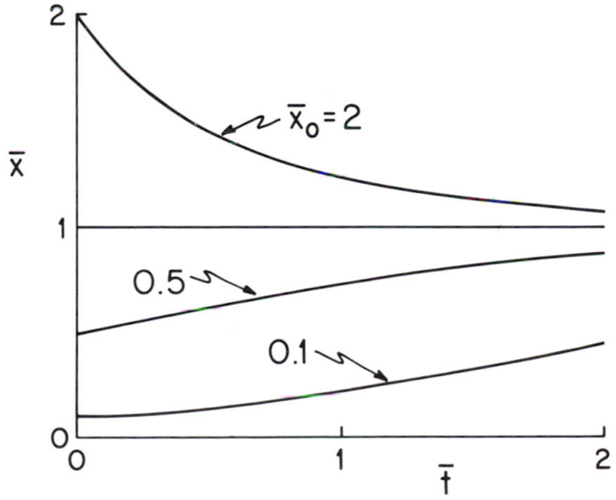

Figure 9.1. Solutions of the logistic equation (9.1) for several initial conditions $\bar{x}_0$. All solutions converge to the fixed point $\bar{x} = 1$.

The solution is

$$\bar{x} = \bar{x}_0 \, e^{\bar{t}} \tag{9.7}$$

where $\bar{x}_0$ is the value of $\bar{x}$ at $\bar{t} = 0$ and is assumed to be finite but small. Thus solutions in the immediate vicinity of the fixed point $\bar{x}_p = 0$ diverge with time and are unstable; this fixed point is unstable.

In order to examine the fixed point $\bar{x}_p = 1$ it is appropriate to introduce the new variable $\bar{x}_1$:

$$\bar{x} = 1 + \bar{x}_1 \tag{9.8}$$

Substitution into (9.3) gives

$$\frac{d\bar{x}_1}{dt} = -\bar{x}_1(1 + \bar{x}_1) \tag{9.9}$$

In the vicinity of the fixed point $\bar{x}_1 = 0$ it is appropriate to neglect the quadratic term in (9.9) with the result

$$\frac{d\bar{x}_1}{dt} = -\bar{x}_1 \tag{9.10}$$

which has the solution

$$\bar{x}_1 = \bar{x}_{10} \, e^{-\bar{t}} \tag{9.11}$$

where $\bar{x}_{10}$ is again assumed to be small but finite. As time evolves $\bar{x}_1$ approaches zero. Thus the solutions in the immediate vicinity of the fixed point $\bar{x}_p = 1$ are stable. The stability of the fixed point $\bar{x}_p = 1$ is clearly illustrated in Figure 9.1. For all initial conditions $\bar{x}_0$ the solutions 'flow' in time towards the stable fixed point $\bar{x}_p = 1$. Also, adjacent solutions tend to converge towards each other. These solutions are not chaotic.

In order to discuss further the behavior of nonlinear equations we consider the van der Pol equation

$$M\frac{d^2x}{dt^2} - \alpha\frac{dx}{dt} + \frac{\beta}{3}\left(\frac{dx}{dt}\right)^3 + kx = 0 \tag{9.12}$$

If $\alpha = \beta = 0$ this is the equation of motion of a spring–mass oscillator system with M the mass, k the spring constant, and x the extension of the spring. The coefficients α and β represent the linear and nonlinear damping terms respectively. Again it is appropriate to introduce nondimensional variables. The frequency of the harmonic oscillator $\omega = (k/M)^{1/2}$ introduces a natural time to the problem. The relative amplitudes of the damping terms α and β introduce a natural length scale $(\alpha/\beta)^{1/2}/\omega$ to the problem. Using these time and length scales we introduce the nondimensional variables $\bar{t}$ and $\bar{x}$; we also define the nondimensional parameter ε according to

$$\bar{t} = \left(\frac{k}{M}\right)^{1/2} t \qquad \bar{x} = \left(\frac{\beta k}{\alpha M}\right)^{1/2} x \qquad \varepsilon = \frac{\alpha}{(kM)^{1/2}} \tag{9.13}$$

Substitution of (9.13) into (9.12) gives

$$\frac{d^2\bar{x}}{d\bar{t}^2} - \varepsilon\left[\frac{d\bar{x}}{d\bar{t}} - \frac{1}{3}\left(\frac{d\bar{x}}{d\bar{t}}\right)^3\right] + \bar{x} = 0 \tag{9.14}$$

with ε the only parameter governing the behavior. In considering the solutions of this second-order nonlinear equation it is standard practice to introduce the definition of velocity

$$\frac{d\bar{x}}{d\bar{t}} = \bar{v} \tag{9.15}$$

and to rewrite (9.14) as

$$\frac{d\bar{v}}{d\bar{t}} = -\bar{x} + \varepsilon(\bar{v} - \tfrac{1}{3}\bar{v}^3) \tag{9.16}$$

Dividing (9.16) by (9.15) gives

$$\frac{d\bar{v}}{d\bar{x}} = \frac{-\bar{x} + \varepsilon(\bar{v} - \tfrac{1}{3}\bar{v}^3)}{\bar{v}} \tag{9.17}$$

The $\bar{v}\bar{x}$-space is known as the phase space or phase plane. Solutions of (9.17) follow phase trajectories in this two-dimensional plane with time $\bar{t}$ as a parameter.

We first consider the solution of (9.17) when $\varepsilon = 0$. In this case it becomes

$$\frac{d\bar{v}}{d\bar{x}} = -\frac{\bar{x}}{\bar{v}} \tag{9.18}$$

which integrates to give

$$\bar{x}^2 + \bar{v}^2 = \bar{x}_0^2 \tag{9.19}$$

Simple harmonic motion is a circle in the phase plane. The radius of the circle is determined by the initial nondimensional amplitude $\bar{x}_0$ (or the initial velocity). The relation (9.19) also represents conservation of energy in this nondissipative system; it is the sum of the potential and kinetic energies. For this system the fixed point at $\bar{x} = \bar{y} = 0$ is known as a center. The behavior is illustrated in Figure 9.2(a).

For finite values of ε it is necessary to solve (9.17) numerically. The result for $\varepsilon = 1$ is given in Figure 9.2(b). Solutions for all initial conditions converge towards a limit cycle; this limit cycle is independent of the initial conditions. The physical reason for this behavior can be seen in the original van der Pol equation (9.12). For small amplitudes the negative linear damping term dominates and the amplitude increases. For large amplitudes the positive cubic damping term dominates and the amplitude decreases. The result is that all solutions converge on the same limit cycle at large times. Many sets of equations that produce deterministic chaos for a range of parameter values produce limit cycles for other parameter values.

Before considering chaotic solutions we will present some further introductory material on singular points. Consider the pair of linear total differential equations

$$\frac{dy}{dt} = ay + bx \tag{9.20}$$

$$\frac{dx}{dt} = cy + fx \tag{9.21}$$

where a, b, c, and f are constants. Dividing (9.20) by (9.21) gives

$$\frac{dy}{dx} = \frac{ay + bx}{cy + fx} \tag{9.22}$$

If $b = -c$ and $a = f = 0$ this reduces to (9.18); the solution is given by (9.19) and is a circle in the xy-plane as illustrated in Figure 9.2(a). If $\alpha = a/f$ and $b = c = 0$ we can write (9.22) as

Figure 9.2. (a) Solution of the nondimensional van der Pol equation (9.14) in the phase plane with $\varepsilon = 0$. The solution is a circle representing simple harmonic motion. The position of the circle is dependent upon initial conditions. (b) Solution of the van der Pol equation (9.14) in the phase plane with $\varepsilon = 1$. The solutions approach a limit cycle independent of initial conditions. Solutions for initial conditions inside the limit cycle spiral out to it and solutions for initial conditions outside the limit cycle spiral into it.

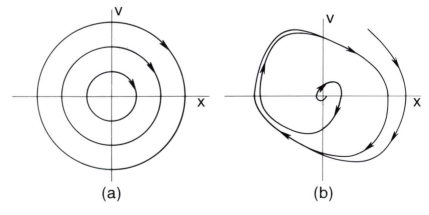

(a)   (b)

$$\frac{dy}{dx} = \frac{\alpha y}{x} \tag{9.23}$$

This equation has the solution

$$y = \gamma x^{\alpha} \tag{9.24}$$

where γ is the constant of integration. If $\alpha > 0$ the fixed point $y = x = 0$ is a node. The behavior for $\alpha = 1$ is illustrated in Figure 9.3(a) and for $\alpha = 2$ in Figure 9.3(b). If $\alpha > 0$ and $f < 0$ the solutions converge on $y = x = 0$ and the fixed point is a stable node. If $\alpha > 0$ and $f > 0$ the solutions diverge from $y = x = 0$ and the fixed point is an unstable node.

If $\alpha < 0$ in (9.23) and (9.24) the fixed point $y = x = 0$ is a saddle point. Its behavior for $\alpha = -1$ is illustrated in Figure 9.3(c). Only the singular solutions $y = 0$ or $x = 0$ enter or leave the fixed point $y = x = 0$. If $x = 0$ we have

$$y = y_0 e^{at} \tag{9.25}$$

and the fixed point is stable for $a < 0$. If $y = 0$ we have

$$x = x_0 e^{ft} \tag{9.26}$$

and the fixed point is stable for $f < 0$. Since a and f must have opposite signs one singular solution will be stable and the other singular solution will be unstable.

We next substitute $b = 1$, $c = -1$, and $a = f = \alpha$ in (9.22) with the result

$$\frac{dy}{dx} = \frac{x + \alpha y}{\alpha x - y} \tag{9.27}$$

Changing to polar coordinates ρ and θ we substitute the variables

$$x = \rho \cos \theta \tag{9.28}$$

$$y = \rho \sin \theta \tag{9.29}$$

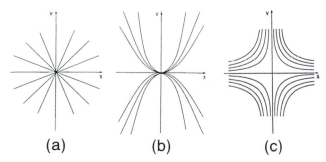

Figure 9.3. Illustration of singular point behavior; (a) and (b) are nodal points and (c) is a saddle point.

(a)　　　(b)　　　(c)

giving

$$\frac{d\rho}{d\theta} = \alpha\rho \tag{9.30}$$

With $\rho = \rho_0$ at $\theta = 0$ this is integrated to give

$$\rho = \rho_0 e^{\alpha\theta} \tag{9.31}$$

This solution is a logarithmic spiral and the fixed point at $y = x = 0$ in (9.27) is known as a spiral.

We now turn to the subject of bifurcations. Solutions to a set of nonlinear equations generally experience a series of bifurcations as they approach chaotic behavior. These bifurcations occur when a parameter of the system is varied. We first consider the equation

$$\frac{dx}{dt} = \mu - x^2 \tag{9.32}$$

where μ is considered to be a parameter. The fixed points of this equation obtained by setting $dx/dt = 0$ are

$$x = \pm\mu^{1/2} \tag{9.33}$$

When μ is negative there are no real fixed points and when μ is positive there are two real fixed points. The transition at $\mu = 0$ from no solutions to two solutions is known as a turning point bifurcation. We examine the stability of the two real roots by linearization. We substitute

$$x = \pm\mu^{1/2} \tag{9.34}$$

into (9.32); after dropping the quadratic term in x_1 we have

$$\frac{dx_1}{dt} = \mp 2\mu^{1/2}x_1 \tag{9.35}$$

Thus the fixed point $x = \mu^{1/2}$ is stable: solutions as they evolve in time converge to it. The fixed point $x = -\mu^{1/2}$ is unstable: solutions as they evolve in time diverge from it. The corresponding bifurcation diagram is given in Figure 9.4(a). For $x < -\mu^{1/2}$ all solutions diverge to $x = -\infty$. For $-\mu^{1/2} < x < +\infty$ all solutions converge to the stable fixed point $x = \mu^{1/2}$.

We next consider the equation

$$\frac{dx}{dt} = x(\mu - x^2) \tag{9.36}$$

where μ is again considered to be a parameter. The fixed points of this equation obtained by setting $dx/dt = 0$ are

$$x = \begin{cases} 0 \\ \pm \mu^{1/2} \end{cases}$$ (9.37)

When μ is negative there is a single real fixed point $x = 0$ and when μ is positive there are three fixed points $x = 0, \pm\mu^{1/2}$. The transition at $\mu = 0$ from one to three solutions is known, for obvious reasons, as a pitchfork bifurcation. A stability analysis shows that for $\mu < 0$ the solution $x = 0$ is stable. For $\mu > 0$ this solution is unstable but the other solutions are stable. The corresponding bifurcation diagram is shown in Figure 9.4(b). For $\mu < 0$ all solutions converge to the stable fixed point $x = 0$. For $\mu > 0$ all solutions for $x > 0$ converge to the stable fixed point $x = \mu^{1/2}$ and all solutions for $x < 0$ converge to the stable fixed point $x = -\mu^{1/2}$.

Finally we consider the pair of equations

$$\frac{dx}{dt} = -\gamma y + [\mu - (x^2 + y^2)]x$$ (9.38)

$$\frac{dy}{dt} = \gamma x + [\mu - (x^2 + y^2)]y$$ (9.39)

As in (9.28) and (9.29) it is again appropriate to introduce polar coordinates ρ and θ in the xy phase plane:

$$x = \rho \cos \theta$$ (9.40)
$$y = \rho \sin \theta$$ (9.41)

Substitution into (9.38) and (9.39) gives

$$\frac{d\theta}{dt} = \gamma$$ (9.42)

$$\frac{d\rho}{dt} = \rho(\mu - \rho^2)$$ (9.43)

Figure 9.4. (a) Illustration of a turning point bifurcation occurring at $\mu = 0$. The stable and unstable fixed points of (9.32) are given as a function of μ. (b) Illustration of a pitchfork bifurcation occurring at $\mu = 0$. The stable and unstable fixed points of (9.36) are given. The transition is from a single stable branch for $\mu < 0$ to three branches, two stable and one unstable, for $\mu > 0$.

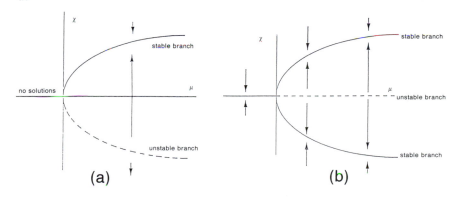

These equations have the fixed point solution $\rho = 0$ ($x = y = 0$); it is stable for $\mu < 0$ and unstable for $\mu > 0$. In addition, for $\mu > 0$, solutions of (9.42) and (9.43) converge to a circular limit cycle given by

$$\rho = \mu^{1/2} \tag{9.44}$$

These solutions are illustrated in Figure 9.5. For $\mu < 0$, all solutions spiral into the stable fixed point $\rho = 0$. For $\mu > 0$, solutions for $\rho > \mu^{1/2}$ spiral into the circular limit cycle given by (9.44): solutions for $\rho < \mu^{1/2}$ spiral outward to this circular limit cycle. The transition from a stable branch for $\mu < 0$ to a stable limit cycle for $\mu > 0$ is a Hopf bifurcation. The van der Pol equation (9.14) also undergoes a Hopf bifurcation at $\varepsilon = 0$.

Figure 9.5. Illustration of a Hopf bifurcation at $\mu = 0$. The limiting solutions of (9.38) and (9.39) are given for various values of μ. For $\mu < 0$ the origin $x = y = 0$ is a stable branch. For $\mu > 0$ there are stable limit cycles which are circles in the xy-plane with radius $\rho = \mu^{1/2}$.

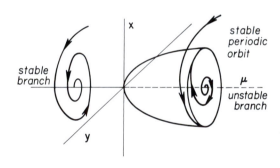

Problems

Problem 9.1. For $b = c = 0$ in (9.20) and (9.21) solve for $y(t)$ and $x(t)$ directly. Show that these solutions reduce to (9.24).

Problem 9.2. Derive (9.30) from (9.27)–(9.29).

Problem 9.3. Solve (9.36) in the vicinity of the three fixed points.

Problem 9.4. Derive (9.42) and (9.43) from (9.38) and (9.39).

Problem 9.5. Consider the equation

$$\frac{dx}{dt} = (1 - x^2)$$

(a) What are the fixed points?

(b) Are they stable or unstable?

(c) Show that $x = \dfrac{(1 + x_0)e^{2t} - 1 + x_0}{(1 + x_0)e^{2t} + 1 - x_0}$ is a solution if $x = x_0$ at $t = 0$.

(d) Sketch solutions for $x_0 = -2, 0, 2$ and discuss in terms of the fixed points.

Problem 9.6. Consider the nondimensional logistic equation (9.3). Determine the solution in the $\bar{v}\bar{x}$ phase plane, where $\bar{v} = \dfrac{dx}{dt}$.

CHAPTER TEN

Logistic map

The concept of deterministic chaos is a major revolution in continuum mechanics (Bergé *et al.*, 1986). Its implications may turn out to be equivalent to the impact of quantum mechanics on atomic and molecular physics. Solutions to problems in solid and fluid mechanics have generally been thought to be deterministic. If initial and boundary conditions on a region are specified then the time evolution of the solution is completely determined. This is in fact the case for linear equations such as the Laplace equation, the heat conduction equation, and the wave equation.

However, the problem of fluid turbulence has remained one of the major unsolved problems in physics. Turbulent flows govern the behavior of the oceans and atmosphere. The appropriate Navier–Stokes equations can be written down but solutions yielding fully developed turbulence cannot be obtained. It is necessary to treat turbulent flows statistically and to carry out spectral analyses.

The concept of deterministic chaos bridges the gap between stable deterministic solutions to equations and deterministic solutions that are unstable to infinitesimal disturbances. Chaotic solutions must also be treated statistically; they evolve in time with exponential sensitivity to initial conditions. A deterministic solution is defined to be chaotic if two solutions that initially differ by a small amount diverge exponentially as they evolve in time. The evolving solutions are predictable only in a statistical sense. A necessary condition that a solution be chaotic is that the governing equations be nonlinear.

As our first example of deterministic chaos we consider the logistic map

$$x_{n+1} = ax_n(1 - x_n) \tag{10.1}$$

This is a recursive relation that determines the sequence of values $x_0, x_1, x_2, \ldots$. An initial value, x_0, is chosen; this value is substituted into (10.1) as x_n and x_1 is determined as x_{n+1}. This value of x_1 is then substituted as x_n and x_2 is determined as x_{n+1}. The process is continued iteratively. This is referred to as a map because the algebraic relation maps out a sequence of values of x_n; $x_0, x_1, x_2, \ldots$. The procedure is best illustrated by taking a specific example. Asume that $a = 1$ and $x_0 = 0.5$ and substitute these values into (10.1), giving $x_1 = 0.25$. This value is then substituted as x_n and we find that $x_2 = 0.1875$. The iterations of (10.1) were studied by May (1976) and have a remarkable range of behavior depending upon the value of a that is chosen. This is in striking contrast to the rather dull behavior of the logistic differential equation given in (9.1).

In order to better understand the behavior of the logistic map it is appropriate to study the functional relation

$$f(x) = ax(1 - x) \tag{10.2}$$

The fixed points x_f of this equation are obtained by setting $f(x_f) = x_f$ with the result

$$x_f = ax_f(1 - x_f) \tag{10.3}$$

This is equivalent to setting $x_{n+1} = x_n$. The two fixed points obtained by solving (10.3) are

$$x_f = 0 \tag{10.4}$$

$$x_f = 1 - \frac{1}{a} \tag{10.5}$$

An essential question is whether the iterative mapping given by (10.1) will evolve to these fixed points. Depending on the behavior of $f(x)$ in the vicinity of the fixed point the fixed point can be either stable or unstable. Solutions will iterate towards stable fixed points and will iterate away from unstable fixed points. We introduce

$$\Gamma = \left(\frac{df}{dx} \right)_{x=x_f} \tag{10.6}$$

This is the slope of the function $f(x)$ evaluated at the fixed point x_f. If $|\Gamma| < 1$, where $|\Gamma|$ is the absolute value of Γ, the fixed point is attracting (stable), but if $|\Gamma| > 1$ the fixed point is repelling (unstable). For the logistic map from (10.2) we find that

$$\Gamma = a \quad \text{at} \quad x_f = 0 \tag{10.7}$$

$$\Gamma = 2 - a \quad \text{at} \quad x_f = 1 - \frac{1}{a} \tag{10.8}$$

For positive values of a we find that the fixed point at $x_f = 0$ is stable for $0 < a < 1$ and unstable for $a > 1$. The fixed point $x_f = 1 - a^{-1}$ is unstable for $0 < a < 1$, stable for $1 < a < 3$, and unstable for $a > 3$.

We next examine a sequence of iterations of the logistic map (10.1). As our first example we consider the iteration for $a = 0.8$ as illustrated in Figure 10.1. The curve represents the function $f(x)$ given by (10.2) for $a = 0.8$. Taking $x_0 = 0.5$ we draw a vertical line; its intersection with the parabolic curve gives $x_1 = 0.2$. A horizontal line drawn from this intersection to the diagonal line of unit slope transfers x_{n+1} to x_n. A vertical line is drawn to the parabola giving $x_2 = 0.128$. Further iterations give $x_3 = 0.0892928$, $x_4 = 0.06505567$, etc. The sequence iterates to the stable fixed point $x_f = 0$. All iterations converge to $x_f = 0$ for $0 < x_0 < 1$. As our next example we consider two iterations for $a = 2.5$ as illustrated in Figure 10.2. The parabolic curve $f(x)$ given by (10.2) now intersects the diagonal at the two fixed points given by (10.4) and (10.5), $x_f = 0$ and $x_f = 0.6$. For $x_f = 0$, $\lambda = 2.5$ and the singular point is unstable; for $x_f = 0.6$, $\lambda = 0.5$ and the singular point is stable. For $x_0 = 0.1$ we find $x_1 = 0.225$, $x_2 = 0.43594$, $x_3 = 0.61474$ and the iteration converges on the fixed point $x_f = 0.6$.

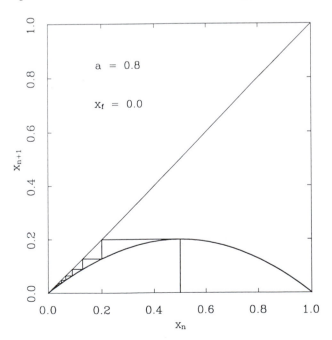

Figure 10.1. Illustration of the iteration of the logistic map (10.1) for $a = 0.8$. The iteration from $x_0 = 0.5$ converges on the stable fixed point $x_f = 0$.

For $x_0 = 0.8$ we find $x_1 = 0.4$ and $x_2 = 0.6$ with no further iteration required. All iterations converge to $x_f = 0.6$ for $0 < x_0 < 1$. This is consistent with the stability of the fixed points discussed above.

At $a = 3$ a flip bifurcation occurs. Both singular points are unstable and the iteration converges on a limit cycle oscillating between x_{f1} and x_{f2}. The period of the oscillation doubles from one iteration, $n = 1$, for $a < 3$ to two iterations, $n = 2$, for $a > 3$. The values of x_{f1} and x_{f2} are obtained from the logistic map (10.1) by writing

$$x_{f2} = ax_{f1}(1 - x_{f1}) \tag{10.9}$$

$$x_{f1} = ax_{f2}(1 - x_{f2}) \tag{10.10}$$

The limit cycle oscillates between x_{f1} and x_{f2}. As an example of the period $n = 2$ limit cycle we consider the iteration for $a = 3.1$ given in Figure 10.3. The iteration from $x_0 = 0.1$ approaches the limit cycle that oscillates between $x_{f1} = 0.558$ and $x_{f2} = 0.765$. The $n = 2$ limit cycle occurs in the range $3 < a < 3.449479$. At $a = 3.449479$ another flip bifurcation occurs and the period doubles again so that $n = 4$. As an example of the period $n = 4$ limit cycle we consider the iteration for $a = 3.47$, which is given in Figure 10.4. The iteration from $x_0 = 0.1$ approaches the $n = 4$ limit cycle that oscillates between $x_{f1} = 0.403, x_{f3} = 0.835, x_{f2} = 0.479$ and $x_{f4} = 0.866$. The $n = 4$ limit

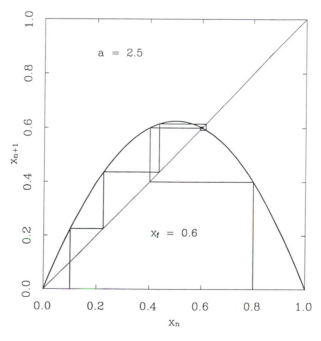

Figure 10.2. Illustration of the iteration of the logistic map (10.1) for $a = 2.5$. The iterations from $x_0 = 0.1$ and $x_0 = 0.8$ converge on the stable fixed point at $x_f = 0.6$.

Figure 10.3. Illustration of
the iteration of the logistic
map (10.1) for $a = 3.1$. The
iteration from $x_0 = 0.1$
converges to the $n = 2$ limit
cycle between $x_{f1} = 0.558$
and $x_{f2} = 0.765$.

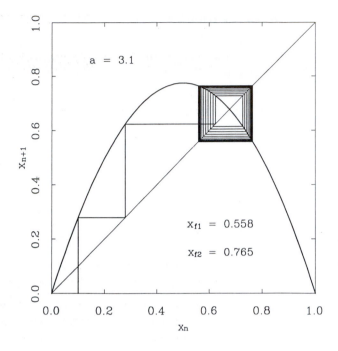

Figure 10.4. Illustration of
the iteration of the logistic
map (10.1) for $a = 3.47$. The
iteration from $x_0 = 0.1$
converges to the $n = 4$ limit
cycle between $x_{f1} = 0.403$,
$x_{f3} = 0.835$, $x_{f2} = 0.479$,
and $x_{f4} = 0.866$.

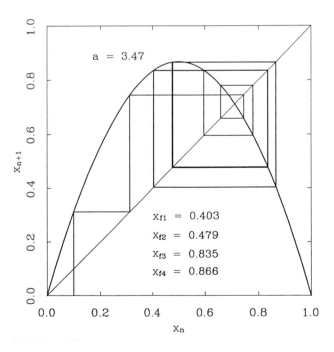

cycle occurs in the range $3.449499 < a < 3.544090$. At larger values of a higher-order limit cycles are found. They are summarized as follows:

$3 < a < 3.449499$	$n = 2$	$k = 1$
$3.449499 < a < 3.544090$	$n = 4$	$k = 2$
$3.544090 < a < 3.564407$	$n = 8$	$k = 3$
$3.564407 < a < 3.568759$	$n = 16$	$k = 4$
$3.568759 < a < 3.569692$	$n = 32$	$k = 5$
$3.569692 < a < 3.569891$	$n = 64$	$k = 6$
$3.569891 < a < 3.569934$	$n = 128$	$k = 7$
$\vdots$	$\vdots$	

where n is the period of the limit cycle and k is the number of flip bifurcations that have occurred. Period-doubling flip bifurcations occur at a sequence of values a_k, where $a_1 = 3$, $a_2 = 3.449499$, $a_3 = 3.544090$, $a_4 = 3.564407$, $a_5 = 3.568759$, $a_6 = 3.569692$, $a_7 = 3.569891$, $a_8 = 3.569934$, etc. In the region $3.569946 < a < 4$ windows of chaos and multiple cycles occur.

The values of a_k approximately satisfy the Feigenbaum relation

$$a_k - a_{k-1} = F(a_{k+1} - a_k) \tag{10.11}$$

where $F = 4.669202$ is the Feigenbaum constant. This becomes a better approximation as k becomes larger. This relation indicates a fractal-like, scale-invariant behavior for the period-doubling sequence of bifurcations. The Feigenbaum relation can also be written in the form

$$a_\infty = \frac{1}{F - 1}(Fa_{k+1} - a_k) \tag{10.12}$$

Thus the initial values of the period-doubling sequence can be used to predict the onset of chaotic behavior at a_∞. Taking $a_1 = 3$ and $a_2 = 3.449499$ we find that $a_\infty = 3.572005$ from (10.12). Taking a_2 and $a_3 = 3.544090$ we find $a_\infty = 3.569870$. Taking a_3 and $a_4 = 3.564407$ we find $a_\infty = 3.569944$. Taking a_4 and $a_5 = 3.568759$ we find $a_\infty = 3.569945$. These are clearly converging on the observed value of $a_\infty = 3.569946$.

We now turn to the behavior of the logistic map in the region of chaotic behavior. An example illustrating chaotic behavior is given in Figure 10.5 with $a = 3.9$; one thousand iterations are shown and no convergence to a limit cycle is observed. For $a = 4$ the logistic map (10.1) becomes

$$x_{n+1} = 4x_n(1 - x_n) \tag{10.13}$$

This iteration can be expressed analytically by taking

$$x_0 = \sin^2 \pi\beta \quad (0 < \beta < 1) \tag{10.14}$$

Substitution of (10.14) into (10.13) gives

$$x_1 = 4(\sin^2 \pi\beta)(1 - \sin^2 \pi\beta) = 4\sin^2 \pi\beta \cos^2 \pi\beta = \sin^2(2\pi\beta) \tag{10.15}$$

The nth iteration of this relation is

$$x_n = \sin^2 2^n \pi\beta \tag{10.16}$$

Provided β is not a rational number the values of x_n jump around randomly and fully chaotic behavior is obtained.

The route to chaos and the windows of chaotic behavior of the logistic map are illustrated in the bifurcation diagram given in Figure 10.6. The asymptotic, large n, behavior of the map is illustrated for $2.9 < a < 4.0$. At $a = 2.9$ the fixed point $x_f = 0.655$ is shown. At $a = 3$ the fixed point is $x_f = 0.66767$ and the period-doubling flip bifurcation to the $n = 2$ limit cycle is shown. In the interval $3 < a < 3.449499$ the two values of x_f corresponding to the $n = 2$ limit cycle are given. At $a = 3.449499$ the period-doubling flip bifurcation to the $n = 4$ limit cycle is shown. In the interval $3.449499 < a < 3.544090$ the four values of x_f corresponding to the $n = 4$ limit cycle are given. In the interval $3.544090 < a < 3.569946$ an infinite sequence of period-doubling flip bifurcations occurs as

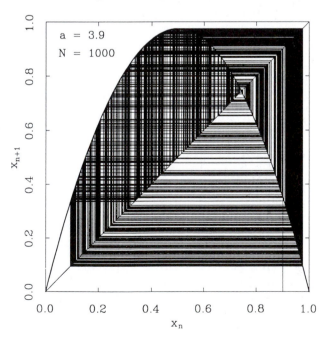

Figure 10.5. Illustration of the iteration of the logistic map (10.1) for $a = 3.9$. The iteration from 0.9 gives chaotic behavior; 1000 iterations are shown.

$n \to \infty$. For the higher values of a the windows of chaotic behavior are illustrated by the cloud of points.

Chaotic behavior results in an infinite set of random values of x_f with a well defined range of values; this range is clearly illustrated in Figures 10.5 and 10.6. The maximum value of x_f is the maximum value of $f(x)$ from (10.2) and this maximum is at $x = 0.5$; thus, we have

$$x_{fmax} = \frac{a}{4} \tag{10.17}$$

Taking $a = 3.9$ we have $x_{fmax} = 0.975$ which is in agreement with the example given in Figure 10.5. The minimum value of x_f is obtained by substituting (10.17) into (10.1) with the result

$$x_{fmin} = \frac{a^2}{4}\left(1 - \frac{a}{4}\right) \tag{10.18}$$

Taking $a = 3.9$ we have $x_{fmin} = 0.0950625$ which is again in agreement with the example given in Figure 10.5.

Chaotic behavior can be quantified in terms of the Lyapunov exponent λ. The definition of the Lyapunov exponent is

$$dx_n = dx_0 2^{\lambda n} \tag{10.19}$$

where dx_n is the incremental difference after the nth iteration if dx_0 is the incremental difference in the initial value. If the Lyapunov

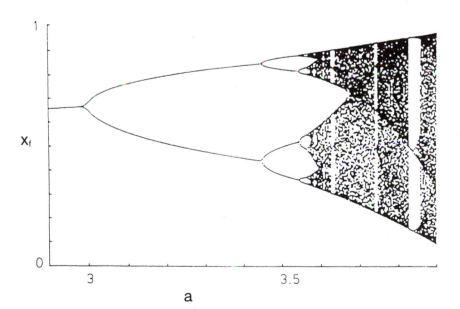

Figure 10.6. Bifurcation diagram showing the asymptotic behavior for large n of the logistic map (10.1) as a function of a.

exponent is negative, adjacent solutions converge and deterministic solutions are obtained. If the Lyapunov exponent is positive, adjacent solutions diverge exponentially and chaos ensues. In order to determine the Lyapunov exponent we consider the incremental divergence in a single iteration by writing (10.1) in the form

$$x_{n+1} + dx_{n+1} = f(x_n + dx_n) = f(x_n) + \left(\frac{df}{dx}\right)_n dx_n \tag{10.20}$$

where $f(x)$ is the functional form of the mapping; for the logistic map it is given by (10.2). Since

$$x_{n+1} = f(x_n) \tag{10.21}$$

by definition, (10.20) can be written

$$dx_{n+1} = \left(\frac{df}{dx}\right)_n dx_n \tag{10.22}$$

And for the logistic map from (10.2) we find

$$dx_{n+1} = a(1 - 2x_n)\,dx_n \tag{10.23}$$

From (10.19) and (10.22) the definition of the Lyapunov exponent is

$$\lambda = \lim_{m\to\infty} \frac{1}{m}\sum_{n=0}^{m} \log_2 \left|\left(\frac{df}{dx}\right)_n\right| \tag{10.24}$$

where $\log_2$ is the logarithm to the base 2. The Lyapunov exponents λ for the logistic map (10.1) are given in Figure 10.7 for a range of

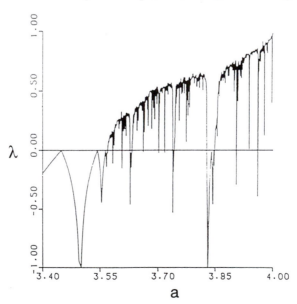

Figure 10.7. Lyapunov exponents λ from (10.24) for the logistic map (10.1) as a function of the parameter a.

values for a. The windows of chaotic behavior for $3.569946 < a < 4$ where λ is positive are clearly illustrated. The Lyapunov exponent goes to zero at each flip bifurcation as shown.

Consider as a particular example the iteration for $a = 4$ given by (10.16). For this case we find

$$dx_n = 2\sin(2^n\pi\beta)\cos(2^n\pi\beta)2^n\pi\,d\beta \tag{10.25}$$

and

$$dx_0 = 2\sin\pi\beta\cos\pi\beta\pi\,d\beta \tag{10.26}$$

Combining (10.19), (10.25), and (10.26) gives

$$2^{\lambda n} = \left[\frac{\sin(2^n\pi\beta)\cos(2^n\pi\beta)}{\sin\pi\beta\cos\pi\beta}\right]2^n \tag{10.27}$$

Although the coefficient is variable, the growth with n as $n \to \infty$ requires that $\lambda = 1$. Thus the Lyapunov exponent for this special case is unity and the iteration is fully chaotic.

Problems

Problem 10.1. Determine x_1, x_2, x_3, and x_4 for the logistic map (10.1) taking $a = 0.5$ and $x_0 = 0.5$. What is the value of x_f?

Problem 10.2. Determine x_1, x_2, x_3, and x_4 for the logistic map (10.1) taking $a = 0.9$ and $x_0 = 0.75$. What is the value of x_f?

Problem 10.3. Determine x_1, x_2, x_3, and x_4 for the logistic map (10.1) taking $a = 2$ and $x_0 = 0.2$. What is that value of x_f?

Problem 10.4. Determine x_1, x_2, x_3, and x_4 for the logistic map (10.1) taking $a = 2.5$ and $x_0 = 0.3$. What is that value of x_f?

Problem 10.5. Determine x_{f1} and x_{f2} for the logistic map (10.1) taking $a = 3.2$.

Problem 10.6. Determine x_{f1} and x_{f2} for the logistic map (10.1) taking $a = 3.4$.

Problem 10.7. For $a = 3.7$ the logistic map (10.1) is fully chaotic. What are the maximum and minimum values of x_f?

Problem 10.8. For $a = 3.8$ the logistic map (10.1) is fully chaotic. What are the maximum and minimum values of x_f?

Problem 10.9. Determine x_0, x_1, x_2, x_3, and x_4 for the logistic map (10.1) taking $a = 4$ and $\beta = (2\pi)^{-1}$.

Problem 10.10. Determine x_0, x_1, x_2, x_3, and x_4 for the logistic map (10.1) taking $a = 4$ and $\beta = (3\pi)^{-1}$.

CHAPTER ELEVEN

Slider-block models

We next turn to a lower-order example of deterministic chaos that has somewhat more direct applications to geology and geophysics. As discussed in Chapter 4 we accept the hypothesis that earthquakes occur repetitively on pre-existing faults. A simple model for the behavior of a fault is a slider block pulled by a spring as illustrated in Figure 11.1 (Burridge and Knopoff, 1967). The block is constrained to move smoothly along the surface. It interacts with the surface through friction; this friction prevents sliding of the block until a critical value of the pulling force is reached. The block sticks and the force in the spring increases until it equals the frictional resistance to sliding on the surface, then slip occurs. The extension of the spring is analogous to the elastic strain in the rock adjacent to a fault. The slip is analogous to an earthquake on a fault. This is stick–slip behavior. The stored elastic strain in the spring is relieved; this is known as elastic rebound on a fault.

The behavior of this simple spring–block model will now be studied quantitatively. A constant velocity driver moving at velocity

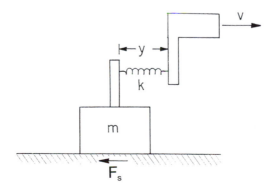

Figure 11.1. Illustration of the slider-block model for fault behavior. The constant velocity driver extends the spring until the force ky exceeds the static friction force F_s.

v extends the spring with spring constant k until the pulling force ky equals the frictional static resisting force F_s. The static condition for the onset of sliding is thus

$$ky = F_s \tag{11.1}$$

Once sliding begins the equation of motion for the block is

$$m\frac{d^2y}{dt^2} + ky = F_d \tag{11.2}$$

where m is the mass of the block and F_d is the dynamic friction. The sliding is analogous to an earthquake and relieves the stress in the spring in analogy to elastic rebound. The further assumption is made that the loading velocity of the driver, v, is sufficiently slow so that we may assume it to be zero during the sliding of the block. This is reasonable since an earthquake only lasts a few tens of seconds whereas the interval between earthquakes on a fault is typically hundreds of years or more.

The static–dynamic friction law is the simplest that generates stick–slip behavior, a necessary and sufficient condition for stick–slip behavior is that the static friction exceeds the dynamic friction, $F_s > F_d$. A variety of empirical velocity-weakening friction laws are in agreement with laboratory observations and also generate stick–slip behavior. Dynamic instabilities associated with complicated friction laws are well known from single-block models (Byerlee, 1978; Dieterich, 1981; Ruina, 1983; Rice and Tse, 1986). Slider-block models have been used to simulate foreshocks, aftershocks, pre- and post-seismic slip, and earthquake statistics (Dieterich, 1972; Rundle and Jackson, 1977; Cohen, 1977; Cao and Aki, 1984, 1986). Gu *et al.* (1984) found some chaotically bounded oscillations; Nussbaum and Ruina (1987) used a two-block model with spatial symmetry and found periodic behavior. Huang and Turcotte (1990b) studied the same system without spatial symmetry and obtained classic chaotic behavior. Carlson and Langer (1989) used many blocks and also obtained chaotic behavior.

We first consider the solution for the behavior of the single block shown in Figure 11.1. It is convenient to introduce the nondimensional variables

$$\phi = \frac{F_s}{F_d} \quad \tau = t\left(\frac{k}{m}\right)^{1/2} \quad Y = \frac{ky}{F_s} \tag{11.3}$$

In terms of these variables the sliding condition (11.1) becomes

$$Y = 1 \tag{11.4}$$

and the equation of motion (11.2) becomes

$$\frac{d^2 Y}{d\tau^2} + Y = \frac{1}{\phi} \tag{11.5}$$

We assume sliding starts at $\tau = 0$ with $Y = 1$ (11.4) and $dY/d\tau = 0$. The applicable solution is

$$Y = \frac{1}{\phi} + \left(1 - \frac{1}{\phi}\right)\cos\tau \tag{11.6}$$

Sliding ends at $\tau = \pi$ when $dY/d\tau$ is again zero. When the velocity is zero the friction jumps to its static value preventing further sliding. The position of the block at the end of sliding is $Y = (2/\phi) - 1$ so that the slip during sliding is

$$\Delta Y = 2\left(\frac{1}{\phi} - 1\right) \tag{11.7}$$

The dependence of Y and $dY/d\tau$ on τ during sliding are given in Figure 11.2 for $\phi = 1.25$. For this case Y drops from 1 to 0.6 during sliding and $\Delta Y = 0.4$. After sliding is completed the spring extends due to the velocity of the driver until Y again equals unity and the cycle repeats. With a single slider block periodic behavior is obtained. The variables Y and dY/dt define a phase plane for the solution.

We next consider the behavior of a pair of slider blocks as illustrated in Figure 11.3. We will show that the behavior of the blocks can be a classical example of deterministic chaos. The blocks are an analog of two interacting faults or two interacting segments of a single fault. A constant velocity driver drags the blocks over the surface at a mean velocity v. The two blocks are coupled to each other and to the constant velocity driver with springs whose constants

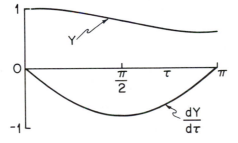

Figure 11.2. Dependence of Y and $dY/d\tau$ on τ during a sliding episode assuming $\phi = 1.25$.

are k_c, k_1, and k_2. Other model parameters are the block masses m_1 and m_2 and the frictional forces F_1 and F_2. The position coordinates of the blocks relative to the constant velocity driver are y_1 and y_2. The static conditions for the onset of sliding are the force balances

$$k_1 y_1 + k_c(y_1 - y_2) = F_{S1} \tag{11.8}$$

$$k_2 y_2 + k_c(y_2 - y_1) = F_{S2} \tag{11.9}$$

Once sliding begins the applicable equations of motion are

$$m_1 \frac{d^2 y_1}{dt^2} + k_1 y_1 + k_c(y_1 - y_2) = F_{D1} \tag{11.10}$$

$$m_2 \frac{d^2 y_2}{dt^2} + k_2 y_2 + k_c(y_2 - y_1) = F_{D2} \tag{11.11}$$

In order to simplify the model we assume $m_1 = m_2 = m$, $k_1 = k_2 = k$, and $F_{S1}/F_{D1} = F_{S2}/F_{D2} = \phi$.

In addition to the nondimensional variables introduced in (11.3) we write

$$\alpha = \frac{k_c}{k}, \quad \beta = \frac{F_{S2}}{F_{S1}} \tag{11.12}$$

The parameter α is a stiffness parameter. If $\alpha = 0$ the blocks are completely decoupled and each will exhibit the periodic behavior described above for a single block. As $\alpha \to \infty$ the blocks become locked together and act as a single block that again exhibits the periodic behavior described above. If $\beta = 1$ there is complete symmetry between the two blocks, $\beta \neq 1$ introduces an asymmetry. In terms of the nondimensional parameters the sliding conditions (11.8) and (11.9) become

$$Y_1 + \alpha(Y_1 - Y_2) = 1 \tag{11.13}$$

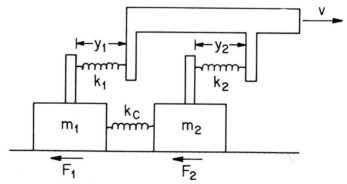

Figure 11.3. Illustration of the two block model. The constant velocity driver extends the springs until sliding of a block commences. In some cases sliding of one block induces the sliding of the second block.

$$Y_2 + \alpha(Y_2 - Y_1) = \beta \qquad (11.14)$$

and the equations of motion (11.10) and (11.11) become

$$\frac{d^2 Y_1}{d\tau^2} + Y_1 + \alpha(Y_1 - Y_2) = \frac{1}{\phi} \qquad (11.15)$$

$$\frac{d^2 Y_2}{d\tau^2} + Y_2 + \alpha(Y_2 - Y_1) = \frac{\beta}{\phi} \qquad (11.16)$$

The blocks are expected to exhibit stick–slip behavior for $\phi > 1$. The first block will begin sliding if (11.13) is satisfied, the second block will begin to slide if (11.14) is satisfied. Together (11.13) and (11.14) define a failure envelope in the $Y_1 Y_2$-plane. The sliding behavior is governed by (11.15) and (11.16). In some cases the sliding of one block induces the sliding of the second block.

The solutions for the behavior of the blocks can be represented in a four-dimensional phase plane consisting of Y_1, Y_2, $dY_1/d\tau$, and $dY_2/d\tau$. For simplicity we consider the projection of the solution onto the $Y_1 Y_2$ plane.

We first consider the symmetric case in which both blocks have the same frictional behavior, i.e. $\beta = 1$. An example with $\alpha = 3$ and $\phi = 1.25$ is given in Figure 11.4. The diagonal lines converging at $Y_1 = Y_2 = 1$ are the failure envelope given by (11.13) and (11.14). A periodic orbit is given by *abcd* in Figure 11.4. At point *a* block 2 fails with $Y_1 = 0.780$ and $Y_2 = 0.835$. During the slip of block 2, block 1 remains fixed and the slip of block 2 is represented by the vertical line *ab* in the $Y_1 Y_2$ phase plane. The termination of the sliding of block 2 is obtained from (11.16). Sliding of block 2 terminates at point *b* where $Y_1 = 0.780$ and $Y_2 = 0.735$. From point *b* to point *c* the blocks stick and the springs extend due to the movement of the constant velocity driver. The increments in Y_1 and Y_2 are equal and the strain accumulation phase is represented by the diagonal line *bc* which has unit slope. The termination of this strain occurs when this line intercepts the failure envelope. This occurs at point *c* where $Y_1 = 0.865$ and $Y_2 = 0.820$. During the slip of block 1, block 2 remains fixed and the slip of block 1 is represented by the horizontal line *cd*. The termination of the sliding of block 1 is obtained from (11.15). Sliding of block 1 terminates at point *d* where $Y_1 = 0.765$ and

$Y_2 = 0.820$. Between points d and a the blocks stick and the springs again extend at equal rates due to the movement of the constant velocity driver. The increments in Y_1 and Y_2 are equal and the strain accumulation phase is represented by the diagonal line ad which has unit slope. The termination of this strain accumulation phase occurs when this line intercepts the failure envelope. This occurs at point a and the cycle repeats. The behavior of this symmetrical two-block model is periodic with first one block sliding and then the other.

We next consider an asymmetric case with $\beta = 2.5$, $\alpha = 3.49$, and $\phi = 1.25$. The results are given in Figure 11.5. The behavior is similar to that given in Figure 10.5 and is fully chaotic. The curves that fall outside the failure envelope are cases in which both blocks are sliding simultaneously. When a diagonal strain accumulation line intercepts the upper failure envelope block 2 begins to slide. Because β is relatively large the failure force for block 2 is considerably larger than the failure force for block 1. Thus the vertical failure path for block 2 crosses the failure envelope of block 1 and it begins to slide. The sliding of both blocks results in S-shaped curves. The next strain accumulation phase intercepts the lower failure envelope and block

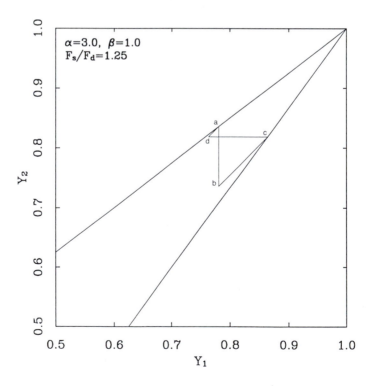

Figure 11.4. Behavior of the symmetrical $\beta = 1$ two-block model illustrated in Figure 11.3 with $\alpha = 3$ and $\phi = 1.25$. Cyclic behavior is obtained with alternating slip of the two blocks.

1 begins to slide. Because of the large force required to induce the failure of block 2 these horizontal failure paths do not cross the upper failure envelope. The result is a chaotic sequence of failures of both blocks together followed by a failure of the weaker block 1.

In order to study this behavior further a bifurcation diagram is given in Figure 11.6. The values of $Y_2 - Y_1$ at the termination of slip are given for various values of α with $\beta = 2.25$ and $\phi = 1.25$. A detailed illustration of the behavior in the range $3.2 < \alpha < 3.5$ is given in Figure 11.7. Solutions evolve to an asymptotic, large-time behavior independent of the initial conditions. As illustrated in Figures 11.6 and 11.7 the system may evolve to limit cycle behavior or chaotic behavior. A series of period-doubling pitchfork bifurcations are clearly illustrated in Figure 11.7. The cyclic behavior for the $n = 2$ limit cycle obtained for $\alpha = 3.25$ is given in Figure 11.8. The cyclic behavior for the $n = 4$ limit cycle obtained for $\alpha = 3.38$ is given in Figure 11.9. These limit cycles evolve into the type of chaotic behavior illustrated in Figure 11.5. The behavior of the asymmetric two-block model is remarkably similar to that of the logistic map.

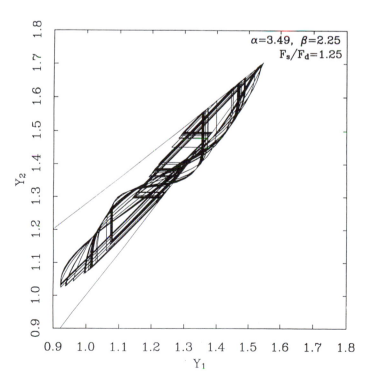

Figure 11.5. Behavior of an asymmetrical two-block model with $\beta = 2.25$, $\alpha = 3.49$ and $\phi = 1.25$. Chaotic behavior is obtained.

Figure 11.6. Bifurcation
diagram for an
asymmetrical two-block
model with $\beta = 2.25$ and
$\phi = 1.25$. The values of
$Y_2 - Y_1$ are given as a
function of α. Singular point,
limit cycle, and chaotic types
of behavior are found.

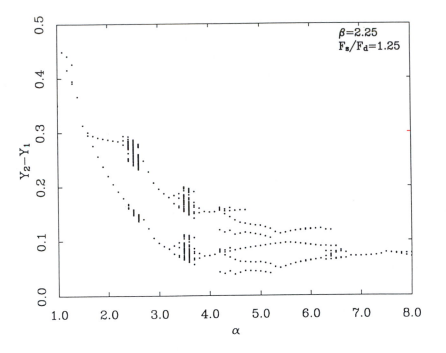

Figure 11.7. Details of the
bifurcation diagram given
in Figure 11.6 for the region
$3.2 < a < 3.5$. The pitchfork
bifurcation illustrated
evolves to deterministic
chaos.

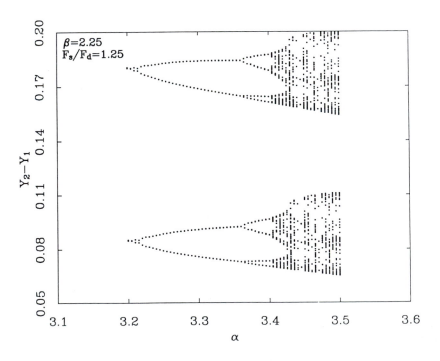

Figure 11.8. Behavior of the asymetric two-block model with $\beta = 2.25$, $\alpha = 3.25$, and $\phi = 1.25$. An $n = 3$ limit cycle is obtained.

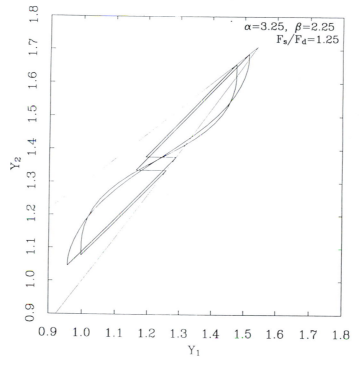

Figure 11.9. Behavior of the asymmetrical two-block model with $\beta = 2.25$, $\alpha = 3.38$, and $\phi = 1.25$. An $n = 4$ limit cycle is obtained.

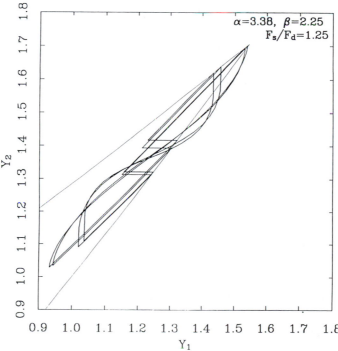

The values of the Lyapunov exponents corresponding to the points given on the bifurcation diagram in Figure 11.6 are given in Figure 11.10. The windows of chaotic behavior are clearly illustrated.

Huang and Turcotte (1990c) have applied the chaotic behavior of the asymmetrical two-block system to two examples of interacting fault segments. The Pacific plate descends beneath the Asian plate resulting in the formation of the Nankai trough along the coast of southwestern Japan. The relative motion between the plates has resulted in a sequence of great earthquakes that have been documented through historical records for the period AD 684–1946. The sequence is marked by an irregular but somewhat repetitive pattern in which whole section failures occur following several alternate failures of single segments. In the two-block model the simultaneous slip of both blocks corresponds to an earthquake that ruptures the entire section and single block failures correspond to an earthquake on a single segment. Taking $\beta = 1.05$ and $\alpha = 0.81$, Huang and Turcotte (1990c) found chaotic model behavior that strongly resembled the observed sequence of earthquakes in the Nankai trough.

Another example is the interaction between the Parkfield segment and the rest of the south central locked segment of the San Andreas fault in California. A sequence of magnitude six earthquakes occurred

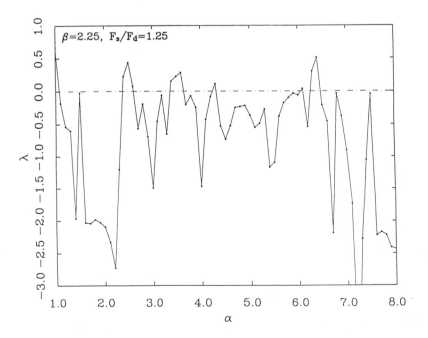

Figure 11.10. Dependence of the Lyapunov exponent on α corresponding to the bifurcation diagram given in Figure 11.6.

on the Parkfield segment in 1881, 1901, 1922, 1934, and 1966. The last great earthquake on the locked segment to the south occurred in 1857 and is also associated with a rupture on the Parkfield segment. Taking $\beta = 2$ and $\alpha = 1.2$, Huang and Turcotte (1990c) found chaotic model behavior similar to that described above. A sequence of slip events on the weaker block often preceded the simultaneous slip of the weaker and stronger blocks. The model simulation suggested two alternative scenarios for a great southern California earthquake following a sequence of Parkfield earthquakes. In the first case a Parkfield earthquake will transfer sufficient stress to trigger the great southern California earthquake; the Parkfield earthquake is thus essentially a foreshock for the great earthquake. In the second case a small additional strain after a Parkfield earthquake will trigger an earthquake on the southern section and this will result in an additional displacement on the Parkfield section. The evolution of the system is chaotic: its evolution is not predictable except in a statistical sense.

Spring-block models are a simple analogy to the behavior of faults in the earth's crust. However, the chaotic behavior of low-dimensional analog systems often indicates that natural systems will also behave chaotically. Thus it is reasonable to conclude that the interaction between faults that leads to the fractal frequency–magnitude statistics discussed in Chapter 4 is an example of deterministic chaos. The prediction of earthquakes is not possible in a deterministic sense. Only a probabilistic approach to the occurrence of earthquakes will be possible.

Problems

Problem 11.1. Consider a single slider block with $\phi = 1.5$. (a) At what value of Y does slip occur? (b) What is the value of Y after slip?

Problem 11.2. Consider a single slider block with $\phi = 3$. (a) At what value of Y does slip occur? (b) What is the value of Y after slip?

Problem 11.3. For a single slider block determine the dependence of $V = dY/dT$ on Y during slip.

Problem 11.4. Consider a pair of slider blocks with $\alpha = 0$, $\beta = 1$ and $\phi = 2$. Assume that initially $Y_1 = 0.5$, $Y_2 = 0$. (a) What are the values of Y_1 and Y_2 when block 1 first slips? (b) What are the values of Y_1 and Y_2 after block 1 slips? (c) What are the values of Y_1 and Y_2 when block 2 first slips? (d) What are the values of Y_1 and Y_2 after block 2 slips? (e) Draw the behavior of the system in the $Y_1 Y_2$ phase plane.

Problem 11.5. Consider a pair of slider blocks with $\alpha = 0$, $\beta = 1$ and $\phi = 4/3$. Assume that initially $Y_1 = 0.75$, $Y_2 = 0.5$. (a) What are the values of Y_1 and Y_2 when block 1 first slips? (b) What are the values of Y_1 and Y_2 after block 1 slips? (c) What are the values of Y_1 and Y_2 when block 2 first slips? (d) What are the values of Y_1 and Y_2 after block 2 slips? (e) Draw the behavior of the system in the $Y_1 Y_2$ phase plane.

CHAPTER TWELVE

Lorenz equations

Sets of coupled nonlinear differential equations can also yield solutions that are examples of deterministic chaos. The classic example is the Lorenz equations. Lorenz (1963) derived a set of three coupled total differential equations as an approximation for thermal convection in a fluid layer heated from below. He showed that the solutions in a particular parameter range had exponential sensitivity to initial conditions and were thus an example of deterministic chaos. This was the first demonstration of chaotic behavior. The Lorenz equations have been studied in detail by Sparrow (1982).

Because of their historical significance and because thermal convection in the earth's mantle drives plate tectonics we will consider the Lorenz equations in some detail. When a fluid is heated its density generally decreases because of thermal expansion. We consider a fluid layer of thickness h that is heated from below and cooled from above; the cool fluid near the upper boundary is dense and the fluid near the lower boundary is light. This situation is gravitationally unstable. The cool fluid tends to sink and the hot fluid tends to rise. This is thermal convection.

Appropriate forms of the continuity, force balance, and energy balance equations are required for a quantitative study of thermal convection. We will restrict our attention to two-dimensional flows in which the velocities are confined to the xy-plane. Continuity of fluid requires that

$$\frac{\partial u}{\partial x} + \frac{\partial v}{\partial y} = 0 \tag{12.1}$$

where u is the horizontal component of velocity in the x-direction and v is the vertical component of velocity in the y-direction, measured downward from the upper boundary. In writing (12.1) the density is assumed to be constant. Force balances in the x and y directions require that

$$\rho\left(\frac{\partial u}{\partial t} + u\frac{\partial u}{\partial x} + v\frac{\partial u}{\partial y}\right) = -\frac{\partial p}{\partial x} + \mu\left(\frac{\partial^2 u}{\partial y^2} + \frac{\partial^2 u}{\partial x^2}\right) \qquad (12.2)$$

$$\rho\left(\frac{\partial v}{\partial t} + v\frac{\partial v}{\partial y} + u\frac{\partial v}{\partial x}\right) = -\frac{\partial p}{\partial y} + \mu\left(\frac{\partial^2 v}{\partial y^2} + \frac{\partial^2 v}{\partial x^2}\right) + \Delta\rho\, g \qquad (12.3)$$

where ρ is density, $\Delta\rho$ the density difference, p pressure, μ viscosity, and g the acceleration due to gravity. The fractional density difference $\Delta\rho/\rho$ is assumed to be small so that $\Delta\rho$ can be neglected except in the buoyancy term $\Delta\rho\, g$ of the vertical force equation (12.3). This is known as the Boussinesq approximation. The terms on the left-hand side of (12.2) and (12.3) are the inertial forces associated with the acceleration of the fluid. The first terms on the right-hand sides of the equations are the pressure forces and the second terms are the viscous forces. The energy balance requires that

$$\rho\, c_p\left(\frac{\partial T}{\partial t} + u\frac{\partial T}{\partial x} + v\frac{\partial T}{\partial y}\right) = k\left(\frac{\partial^2 T}{\partial x^2} + \frac{\partial^2 T}{\partial y^2}\right) \qquad (12.4)$$

where T is temperature, k thermal conductivity, and c_p is the specific heat at constant pressure. The terms on the left-hand side of (12.4) account for the convection of heat and the terms on the right-hand side account for the conduction of heat. A full derivation of these equations has been given by Turcotte and Schubert (1982), pp. 240–74.

In the absence of convection, i.e. $u = v = 0$, the temperature T_0 is a linear conduction profile given by

$$T_c = T_1 + (T_2 - T_1)\frac{y}{h} \qquad (12.5)$$

where T_1 is the constant temperature of the upper boundary ($y = 0$) and T_2 is the constant temperature of the lower boundary ($y = h$). The incompressible continuity equation in two dimensions (12.1) is satisfied if we introduce a stream function ψ defined by

$$u = -\frac{\partial \psi}{\partial y}, \quad v = \frac{\partial \psi}{\partial x} \qquad (12.6)$$

It is also convenient to introduce the temperature difference θ between the actual temperature T and the temperature T_c in the absence of convection:

$$\theta = T - T_c = T - T_1 - (T_2 - T_1)\frac{y}{h} \tag{12.7}$$

The density difference $\Delta\rho$ in the buoyancy term of the vertical force equation (12.3) is related to this temperature difference by

$$\Delta\rho = -\rho\alpha\theta \tag{12.8}$$

where α is the volumetric coefficient of thermal expansion.

Substitution of (12.6)–(12.8) into (12.1)–(12.4) gives

$$\rho\left[\left(\frac{\partial^2}{\partial x^2} + \frac{\partial^2}{\partial y^2}\right)\frac{\partial\psi}{\partial t} + \frac{\partial\psi}{\partial x}\left(\frac{\partial^2}{\partial x^2} + \frac{\partial^2}{\partial y^2}\right)\frac{\partial\psi}{\partial y} - \frac{\partial\psi}{\partial y}\left(\frac{\partial^2}{\partial x^2} + \frac{\partial^2}{\partial y^2}\right)\frac{\partial\psi}{\partial x}\right]$$
$$= \mu\left(\frac{\partial^4\psi}{\partial x^4} + 2\frac{\partial^4\psi}{\partial x^2\partial y^2} + \frac{\partial^4\psi}{\partial y^4}\right) - \rho\alpha g\frac{\partial\theta}{\partial x} \tag{12.9}$$

$$\rho c_p\left[\frac{\partial\theta}{\partial t} - \frac{\partial\psi}{\partial y}\frac{\partial\theta}{\partial x} + \frac{\partial\psi}{\partial x}\frac{(T_2 - T_1)}{h} + \frac{\partial\psi}{\partial x}\frac{\partial\theta}{\partial y}\right] = k\left(\frac{\partial^2}{\partial x^2} + \frac{\partial^2}{\partial y^2}\right)\theta \tag{12.10}$$

The problem has been reduced to the solution of two partial differential equations for ψ and θ.

In order to better understand the roles of various terms it is appropriate to introduce the nondimensional variables

$$\bar{t} = \frac{\kappa t}{h^2}, \quad \bar{x} = \frac{x}{h}, \quad \bar{y} = \frac{y}{h}, \quad \bar{\psi} = \frac{\psi}{\kappa}, \quad \bar{\theta} = \frac{\rho g\alpha h^3\theta}{\kappa\mu} \tag{12.11}$$

where $\kappa = k/(\rho c_p)$ is the thermal diffusivity. With these nondimensional variables two nondimensional parameters govern the behavior of the equations:

$$Ra = \frac{\rho\alpha g(T_2 - T_1)h^3}{\kappa\mu} \tag{12.12}$$

$$Pr = \frac{\mu}{\rho\kappa} \tag{12.13}$$

where Ra is the Rayleigh number and Pr is the Prandtl number. The Rayleigh number is a measure of the strength of the buoyancy forces that drive convection relative to the viscous forces that damp convection. The higher the Rayleigh number the stronger the convection. The Prandtl number is the ratio of the momentum diffusion to the heat diffusion. It is instructive to estimate these two parameters for the earth's mantle. Due to solid-state creep, the earth's mantle has a mean viscosity of around $\mu = 10^{21}$ Pa s, its thickness is $h = 2880$ km, and the temperature increase across it is estimated to be $T_2 - T_1 = 3000$ K. For the rock properties we take $\kappa = 1\,\text{mm}^2\,\text{s}^{-1}$ and $\alpha = 3 \times 10^{-5}\,\text{K}^{-1}$. We assume $g = 10\,\text{m}\,\text{s}^{-2}$ and an average density $\rho = 4000\,\text{kg}\,\text{m}^{-3}$ and find $Ra = 8.6 \times 10^6$ and $Pr = 2.5 \times 10^{23}$, both very large values. The behavior of the earth's mantle will be discussed further in the next chapter.

We now return to the basic equations. Substitution of (12.11) to (12.13) into (12.9) and (12.10) gives

$$\frac{1}{Pr}\left[\left(\frac{\partial^2}{\partial \bar{x}^2} + \frac{\partial^2}{\partial \bar{y}^2}\right)\frac{\partial \bar{\psi}}{\partial \bar{t}} + \frac{\partial \bar{\psi}}{\partial \bar{x}}\left(\frac{\partial^2}{\partial \bar{x}^2} + \frac{\partial^2}{\partial \bar{y}^2}\right)\frac{\partial \bar{\psi}}{\partial \bar{y}} - \frac{\partial \bar{\psi}}{\partial \bar{y}}\left(\frac{\partial^2}{\partial \bar{x}^2} + \frac{\partial^2}{\partial \bar{y}^2}\right)\frac{\partial \bar{\psi}}{\partial \bar{x}}\right]$$

$$= \frac{\partial^4 \bar{\psi}}{\partial \bar{x}^4} + 2\frac{\partial^4 \psi}{\partial \bar{x}^2\,\partial \bar{y}^2} + \frac{\partial^4 \bar{\psi}}{\partial \bar{y}^4} - \frac{\partial \bar{\theta}}{\partial \bar{x}} \quad (12.14)$$

$$\frac{\partial \bar{\theta}}{\partial \bar{t}} - \frac{\partial \bar{\psi}}{\partial \bar{y}}\frac{\partial \bar{\theta}}{\partial \bar{x}} + Ra\frac{\partial \bar{\psi}}{\partial \bar{x}} + \frac{\partial \bar{\psi}}{\partial \bar{x}}\frac{\partial \bar{\theta}}{\partial \bar{y}} = \left(\frac{\partial^2}{\partial \bar{x}^2} + \frac{\partial^2}{\partial \bar{y}^2}\right)\bar{\theta} \quad (12.15)$$

The solution is determined by the two parameters Ra and Pr and the boundary conditions. For small values of the Rayleigh number the viscous forces are sufficiently strong to prevent any flow. Thus there is a critical minimum value of the Rayleigh number for the onset of thermal convection.

We next consider a linearized stability analysis for the onset of convection as given by Rayleigh (1916). Only terms linear in $\bar{\theta}$ and $\bar{\psi}$ are retained and the marginal stability problem is solved by setting $\partial/\partial \bar{t} = 0$. Thus (12.14) and (12.15) become

$$0 = \frac{\partial^4 \bar{\psi}}{\partial \bar{x}^4} + 2\frac{\partial^4 \bar{\psi}}{\partial \bar{x}^2\,\partial \bar{y}^2} + \frac{\partial^4 \bar{\psi}}{\partial \bar{y}^4} - \frac{\partial \bar{\theta}}{\partial \bar{x}} \quad (12.16)$$

$$Ra\frac{\partial \bar{\psi}}{\partial \bar{x}} = \left(\frac{\partial^2}{\partial \bar{x}^2} + \frac{\partial^2}{\partial \bar{y}^2}\right)\bar{\theta} \quad (12.17)$$

It is appropriate to assume solutions that are periodic in the horizontal coordinate $\bar{x}$. Solutions that satisfy (12.17) are

$$\bar{\psi} = \psi_0 \sin\left(\frac{2\pi \bar{x}}{\bar{\lambda}}\right)\sin \pi \bar{y} \quad (12.18)$$

$$\bar{\theta} = \frac{2\bar{\lambda}\psi_0 Ra}{\pi(4 + \bar{\lambda}^2)}\cos\left(\frac{2\pi \bar{x}}{\bar{\lambda}}\right)\sin \pi \bar{y} \quad (12.19)$$

The flow is assumed to be periodic in the horizontal direction with a wavelength λ; the nondimensional wavelength is $\bar{\lambda} = \lambda/h$. The flow consists of counter-rotating cells; each cellular flow has width $\lambda/2$ as illustrated in Figure 12.1. The temperature boundary conditions $\bar{\theta} = 0$ at $\bar{y} = 0$, 1 are satisfied by (12.19). The requirement of no flow through the walls corresponds to $\bar{v} = \partial\bar{\psi}/\partial\bar{x} = 0$ at $\bar{y} = 0$, 1 and these conditions are satisfied by (12.18). If the boundaries of the layer are solid surfaces we would require $\bar{u} = -\partial\bar{\psi}/\partial\bar{y} = 0$ at $\bar{y} = 0$, 1. These are the no-slip conditions requiring that there be no relative motion between a viscous fluid and a bounding solid surface at the solid–fluid interface. If the boundaries of the layer are free surfaces, that is, if there is nothing at the boundaries to exert a sheer stress on the fluid, we would require that the shear stress be zero at $\bar{y} = 0$, 1. These free surface boundary conditions can be written as $\partial\bar{u}/\partial\bar{y} = -\partial^2\bar{\psi}/\partial y^2 = 0$ at $\bar{y} = 0$, 1. The stream function given in (12.18) satisfies the free surface boundary conditions.

In order that (12.18) and (12.19) also satisfy (12.16) we require that

$$Ra = \frac{\pi^4}{4\bar{\lambda}^4}(4 + \bar{\lambda}^2)^3 \tag{12.20}$$

When the nondimensional wavelength $\bar{\lambda}$ is specified this is the minimum value of the Rayleigh number at which convection will occur. As the Rayleigh number for a fluid layer is increased, flow will occur at the wavelength for which the Rayleigh number given by (12.20) is a minimum. This value of the Rayleigh number is given by

$$Ra_c = \frac{27\pi^4}{4} = 657.5 \tag{12.21}$$

This is the critical Rayleigh number for the onset of thermal convection in a fluid layer heated from below. At Rayleigh numbers less than that given by (12.21) thermal convection will not occur. The nondimensional wavelength corresponding to (12.21) is

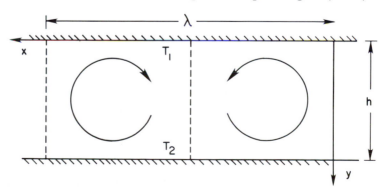

Figure 12.1. Illustration of two-dimensional cellular convection in a fluid layer heated from below.

$$\lambda_c = 8^{1/2} = 2.828 \tag{12.22}$$

This is the wavelength of the initial convective flow that takes the form of counter-rotating, two-dimensional cells. Each cell has a width $2^{1/2}h$, one-half the wavelength of the initial disturbance.

A pitchfork bifurcation occurs at the critical Rayleigh number. If $Ra < Ra_c$ the only solution is the conduction solution, which is stable. Above the critical Rayleigh number the conduction solution remains a solution of the governing equations but it is now unstable. Above the critical Rayleigh number there are two stable convective solutions corresponding to cellular rolls rotating either clockwise or counter-clockwise. This is identical to the pitchfork bifurcation illustrated in Figure 9.4(b).

Because the applicable equations are linear, the stability analysis does not predict the amplitude of the convection. It is not possible to specify the value of ψ_0 in (12.18) and (12.19). In order to determine the amplitude of the thermal convection it is necessary to retain nonlinear terms. One approach to the solution of the full nonlinear equations (12.14) and (12.15) is to expand the variables $\bar\psi$ and $\bar\theta$ in double Fourier series in $\bar x$ and $\bar y$ with coefficients that are functions of time. Lorenz (1963) strongly truncated these series and retained only three terms of the form

$$\bar\psi = \frac{1}{2^{1/2}\bar\lambda}(4 + \bar\lambda^2)A(\tau)\sin\left(\frac{2\pi\bar x}{\bar\lambda}\right)\sin \pi\bar y \tag{12.23}$$

$$\bar\theta = \frac{\pi^3}{4\bar\lambda^4}(4 + \bar\lambda^2)^3\left[\gamma(\tau)\sin 2\pi\bar y - 2^{1/2}B(\tau)\cos\left(\frac{2\pi\bar x}{\bar\lambda}\right)\sin \pi\bar y\right] \tag{12.24}$$

where

$$\tau = \pi^2\left[1 + \left(\frac{2}{\bar\lambda}\right)^2\right]\bar t \tag{12.25}$$

$$r = \frac{Ra}{Ra_c} \tag{12.26}$$

with Ra_c given by (12.20). These equations satisfy the same set of boundary conditions that (12.18) and (12.19) satisfy. The expansion of the stream function (12.23) is essentially identical to the form used in the linear stability analysis (12.18). However, the expansion of the temperature (12.24) includes an additional term that is not dependent on $\bar x$.

It is necessary to derive differential equations for the time dependence of the coefficients $A(\tau)$, $B(\tau)$, $C(\tau)$. This is done by substituting the expansions (12.23) and (12.24) into the governing equations (12.14) and (12.15). Coefficients of the Fourier terms are equated in order to obtain the necessary equations.

All nonlinear terms in the stream function equation (12.14) are neglected. When (12.23) and (12.24) are substituted into this equation the result is

$$\frac{dA}{d\tau} = Pr(B - A) \tag{12.27}$$

As expected this is a linear equation and no further approximations have been made in writing it. In order to maintain a consistent approximation in the energy equation (12.15) it is necessary to retain several nonlinear terms. Substitution of (12.23) and (12.24) into (12.15) gives several terms that are not in the form required. These products of trigonometric functions are reduced to the standard form using the multiple angle formulas

$$\cos(2\pi\bar{x}/\bar{\lambda}) \sin \pi\bar{y} \cos 2\pi\bar{y} = -\cos(2\pi\bar{x}/\bar{\lambda}) \sin \pi\bar{y} \tag{12.28}$$

$$\cos(2\pi\bar{x}/\bar{\lambda}) \sin \pi\bar{y} \cos (2\pi\bar{x}/\bar{\lambda}) \cos \pi\bar{y} = \sin 2\pi\bar{y} \tag{12.29}$$

with higher-order terms neglected.

Equating coefficients gives

$$\frac{dB}{d\tau} = rA - B - AC \tag{12.30}$$

$$\frac{dC}{d\tau} = -bC + AB \tag{12.31}$$

where

$$b = \frac{4}{[1 + (2/\bar{\lambda})^2]} \tag{12.32}$$

The three first-order total differential equations (12.27), (12.30), and (12.31) are the Lorenz equations. These equations would be expected to give accurate solutions to the full equations when the Rayleigh number is slightly supercritical, but large errors would be expected for strong convection because of the extreme truncation.

Solutions of the Lorenz equations represent cellular, two-dimensional convection. Because only one term is retained in the expansion of the stream function, the particle paths are closed and represent streamlines even when the flow is unsteady. The time dependence of the coefficient A determines the velocity of a fluid particle. But the fluid particle follows the same closed trajectory independent of its time variation. The coefficient B represents temperature variations associated with the stream function mode A. The coefficient C represents a horizontally averaged temperature mode. A detailed discussion of the behavior of the Lorenz equations has been given by Sparrow (1982).

In order to examine the behavior of the Lorenz equations we first determine the allowed steady-state solution. Obviously the steady-state solution $A = B = C = 0$ corresponds to heat conduction without flow. An additional pair of allowed solutions is

$$A = B = \pm [b(r-1)]^{1/2} \tag{12.33}$$

$$C = r - 1 \tag{12.34}$$

These solutions correspond to an infinite set of two-dimensional cells as illustrated in Figure 12.1. Adjacent cells rotate in opposite directions; the choice of sign given in (12.33) determines whether a specified cell rotates clockwise or counterclockwise. A stability analysis for the conduction solutions shows that it is stable for $r < 1$ and unstable for $r > 1$. Thus the Lorenz equations exhibit the same type of pitchfork bifurcation at $r = 1$ $(Ra = Ra_c)$ that the full equations do. This is expected since the linearized form of the Lorenz equations is identical to the linearized form of the full equations.

The stability of the steady solution given in (12.33) and (12.34) can also be examined. Expanding about this solution with

$$A = \pm [b(r-1)]^{1/2} + A_1 e^{\lambda \tau} \tag{12.35}$$

$$B = \pm [b(r-1)]^{1/2} + B_1 e^{\lambda \tau} \tag{12.36}$$

$$C = r - 1 + C_1 e^{\lambda \tau} \tag{12.37}$$

and substituting into (12.27), (12.30), and (12.31) with linearization gives the characteristic equation

$$\lambda^3 + (Pr + b + 1)\lambda^2 + (r + Pr)b\lambda + 2b\, Pr\,(r-1) = 0 \tag{12.38}$$

This equation has one real negative root and two complex conjugate roots when $r > 1$. If the product of the coefficients of λ^2 and λ equals the constant term we obtain

$$r = \frac{Pr\,(Pr + b + 3)}{Pr - b - 1} \tag{12.39}$$

At this value of r the complex roots of (12.38) have a transition from negative to positive real parts. This is the critical value of r for the instability of steady convection and represents a subcritical Hopf bifurcation. If $Pr > b + 1$ the steady solutions given by (12.33) and (12.34) are unstable for Rayleigh numbers larger than those given by (12.39).

To examine further the behavior of the Lorenz equations it is necessary to carry out numerical solutions. Following Lorenz (1963) we consider $\bar{\lambda} = 8^{1/2}$, the critical value from (12.22), so that $b = \frac{8}{3}$. For these values the steady-state solution given by (12.23), (12.24), (12.33), and (12.34) becomes

$$\bar{\psi} = \pm [24(r - 1)]^{1/2} \sin\left(\frac{\pi\bar{x}}{2^{1/2}}\right) \sin \pi\bar{y} \tag{12.40}$$

$$\bar{\theta} = \frac{27\pi^3}{4}\left\{(r - 1)\sin 2\pi\bar{y} \mp \left[\frac{16}{3}(r - 1)\right]^{1/2} \cos\left(\frac{\pi\bar{x}}{2^{1/2}}\right) \sin \pi\bar{y}\right\} \tag{12.41}$$

This steady-state solution is valid if $r > 1$ and is less than the critical value given by (12.39).

As the Rayleigh number increases above $r = 1$ the strength of the convection increases, as indicated by (12.40). This results in larger transport of heat by convection and as a result the thermal gradients at the upper and lower boundaries increase. The Nusselt number is a measure of the efficiency of the convective heat transfer across the layer. The Nusselt number Nu is the ratio of the heat transferred by convection to the conductive value without convection. In terms of our nondimensional variables it is given by

$$Nu = 1 + \left\langle\frac{\partial\bar{\theta}}{\partial\bar{y}}\right\rangle_s \frac{1}{Ra} \tag{12.42}$$

where $\langle\ \rangle_s$ indicates an average across either the upper or lower surface. These averages must be equal since the mean heat flux into the layer across the lower boundary must be equal to the mean heat flux out of the layer through the upper boundary. For the steady-state solution given by (12.41) the Nusselt number is

$$Nu = 1 + \frac{2(r - 1)}{r} \tag{12.43}$$

This relation is not in good agreement with experiment when r is significantly larger than unity. This is clearly due to the extreme nature of the truncation, which is only expected to be valid near the stability limit $r = 1$. Nevertheless, it is of interest to explore the behavior of the Lorenz equations for larger values of r.

The critical Rayleigh number for stability of the steady-state solution is $r = 24.74$ for $\bar{\lambda} = 8^{1/2}$ from (12.39). For values of r greater than this, unsteady solutions are expected. A numerical solution of the Lorenz equations for $r = 28$ is given in Figure 12.2. The time dependence of the three variables $A(\tau)$, $B(\tau)$, $C(\tau)$ is obtained. It is convenient to study the solution in the three-dimensional ABC phase space; the time τ is a parameter. The projection of the solution onto the BA-plane is given in Figure 12.2(a) and the projection onto the BC-plane is given in Figure 12.2(b). These are known as phase portraits. The dependence of the variable B on time is given in Figure 12.2(c). The solution randomly oscillates between cellular rolls with clockwise rotation for $B > 0$ and with counterclockwise rotation for $B < 0$. The unstable fixed points from (12.33) and (12.34), $A = B = \pm 72^{1/2}, C = 27$, are the crosses in Figures 12.2(a) and 12.2(b).

This solution is chaotic in that adjacent solutions diverge exponentially in time. The solution oscillates about a fixed point with growing amplitude until it flips into the other quadrant where it oscillates about the other fixed point. The fixed points behave as 'strange attractors'. In Figure 12.2(d) the fixed points are projected onto the rB-plane. The solid lines represent stable fixed points and the dashed lines represent unstable fixed points; the solid circle is a pitchfork bifurcation and the open circles are Hopf bifurcations. The pitchfork bifurcation at $r = 1$ corresponds to the onset of thermal convection; the Hopf bifurcations at $r = 24.74$ correspond to the onset of chaotic flows.

An essential feature of the solution illustrated in Figure 12.2(b) is that the value of C is always positive and oscillates aperiodically around a positive value. It is the C term that gives an approximation to a thermal boundary layer structure. The growth and decay of C around a positive value is an approximation to the growth and separation of the thermal boundary layers at the top and bottom of the convection cell. It is the buoyancy forces in the boundary layers that drive the flow. The direction of the trajectory near $B = 0$ is always toward $C = 0$; the direction of the trajectory for large B is always in the positive C-direction. When C is at its largest the thermal boundary layers are thin, the buoyancy forces are small, and the flow decelerates (B decreases). This quiescence of the convective flow allows the thermal boundary layers to thicken and C decreases. The

Figure 12.2. A numerical solution of the Lorenz equations (12.27), (12.30), (12.31) with $Pr = 10$, $b = 8/3$, $r = 28$. The solution in the ABC phase space is shown projected (a) onto the BA-plane and (b) onto the BC-plane. (c) Time dependence of the coefficient B. (d) The loci of the fixed points are projected onto the rB-plane. The solid lines represent stable fixed points, the broken lines represent unstable fixed points, the solid circle is a pitchfork bifurcation, and the open circles are Hopf bifurcations.

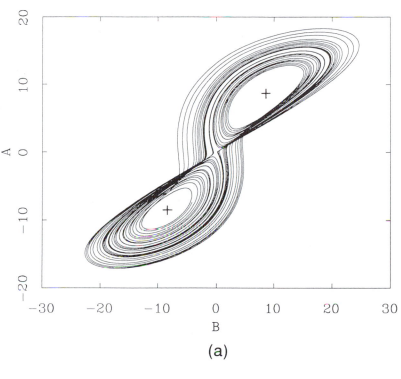

(a)

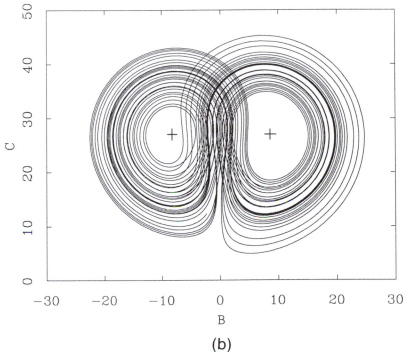

(b)

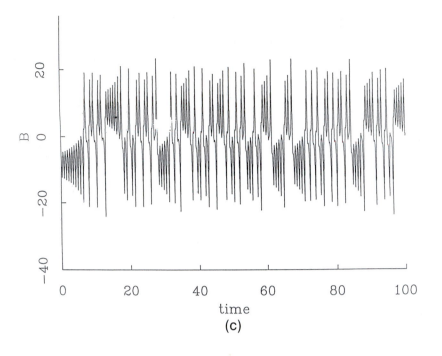

(c)

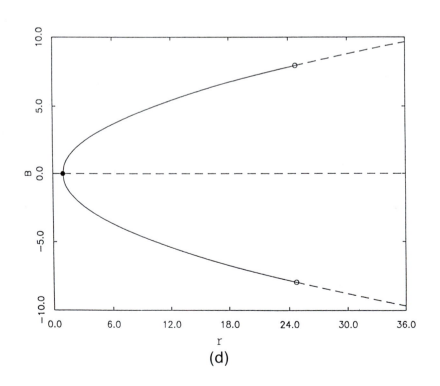

(d)

thickening thermal boundary layers become unstable and the resultant buoyancy accelerates the flow, increasing B; depending on where the instability occurs either clockwise or counterclockwise flow results. As the convective motion increases, the thermal boundary layers again thin, and the value of C grows. The amplitude of the convective motion consequently slows down, and the absolute values of A and B once again move toward zero. Once convective motion has slowed enough, the boundary layer can grow again. Every time the convective motion stops to let the thermal boundary layer grow, the cell is presented with a choice of whether to convect in a clockwise or a counterclockwise direction. It is this freedom of choice in a completely deterministic system that produces the chaotic, aperiodic behavior of the Lorenz attractor.

The essential feature of the solution of the Lorenz equations in this parameter range is deterministic chaos. One consequence of the deterministic chaos of the Lorenz equations is that solutions that begin a small distance apart in phase space diverge exponentially. With essentially infinite sensitivity to initial conditions, the zero-order behavior of a solution is not predictable.

It should be emphasized that the numerical solution of the Lorenz equations given above is not a valid solution for thermal convection in a fluid layer. The Rayleigh number is outside the range of validity of the Lorenz truncation. Nevertheless, this solution has a great significance in that it was the first solution to exhibit the mathematical conditions for chaotic behavior. But the significance goes beyond this. Experimental studies and numerical simulations of thermal convection in a fluid layer heated from below at high Rayleigh numbers and intermediate values of Prandtl number are generally unsteady and 'turbulent'. There is good reason to accept that the solutions to the full set of equations are also chaotic in this range of parameters. Thus weather is taken to be chaotic and deterministic predictions are not possible. It is essential to treat weather as a statistical problem with uncertainties increasing with forward extrapolations.

Problems

Problem 12.1. Show that the critical Rayleigh number given by (12.20) has the minimum value as given by (12.21) and (12.22).

Problem 12.2. For the steady-state solution of the Lorenz equations given in (12.40) and (12.41) determine an expression for the mean horizontal velocities on the boundaries at $\bar{y} = 0, 1$.

CHAPTER THIRTEEN

Is mantle convection chaotic?

The Lorenz equations are a low-order expansion of the full equations applicable to thermal convection in a fluid layer heated from below. For the range of parameters in which chaotic behavior is obtained, the low-order expansion is not valid; higher-order terms should be retained. Nevertheless, the chaotic behavior of the low-order analog is taken as a strong indication that the full equations will also yield chaotic solutions. Numerical solutions of the full equations are strongly time-dependent for high Rayleigh number flows; these solutions appear to be turbulent or chaotic.

It is generally accepted that thermal convection is the primary means of heat transport in the earth's mantle. Heat is produced in the mantle due to the decay of the radioactive isotopes of uranium, thorium, and potassium. Heat is also lost due to the cooling of the earth. The surface plates of plate tectonics are the thermal boundary layers of mantle convection cells. The plates are created by ascending mantle flows at ocean ridges. The plates become gravitationally unstable and founder into the mantle at ocean trenches (subduction zones). Intraplate hot spots such as Hawaii are attributed to mantle plumes that ascend from the hot unstable thermal boundary layer at the base of the convecting mantle.

An important question with regard to the earth is whether mantle convection is chaotic. The earth's solid mantle behaves as a fluid on geological time scales because of thermally activated creep. The discussion in the previous chapter considered only a constant viscosity. This is a poor approximation for the earth's mantle because the dependence of strain on stress is almost certainly nonlinear and is an exponential function of temperature and pressure. Also, the

151

Boussinesq approximation is not applicable because of the significant increase in density with depth (i.e. pressure). Nevertheless, calculations assuming a linear stress–strain relation and constant fluid properties can provide important insights. In the previous chapter we estimated that the Rayleigh number for mantle convection is near 10^7 and the Pandtl number is larger than 10^{23}. The latter is such a large value that it is appropriate to assume that the mantle has an infinite Prandtl number. Because the Prandtl number for the mantle is so large, the momentum terms on the left-hand side of the momentum equations (12.2) and (12.3) can be neglected. Thus the only nonlinear terms are those in the energy equation (12.4). The question is whether these terms can generate chaotic behavior and thermal turbulence.

The first question we address is whether the Lorenz equations yield chaotic solutions in the limit $Pr \rightarrow \infty$. In this limit (12.27) requires

$$A = B \tag{13.1}$$

The substitution of this result into (12.30) and (12.31) gives

$$\frac{\mathrm{d}B}{\mathrm{d}\tau} = (r - 1)B - BC \tag{13.2}$$

$$\frac{\mathrm{d}C}{\mathrm{d}\tau} = -bC + B^2 \tag{13.3}$$

Again these equations have the steady state, $\mathrm{d}/\mathrm{d}\tau = 0$, solution $A = B = C = 0$ corresponding to conduction. These equations also have the same fixed points as the Lorenz equations; these fixed points are given in (12.33) and (12.34) and correspond to cellular rolls rotating either in the clockwise or counterclockwise directions. A stability analysis again shows that the conduction solution is stable for $r < 1$ and unstable for $r > 1$. However, the steady solution consisting of cellular rolls is now stable for the entire range $r > 1$. In the limit of infinite Prandtl number the Lorenz equations do not yield chaotic solutions. This has been taken as evidence that mantle convection is not chaotic.

In order to consider this problem further Stewart and Turcotte (1989) considered a higher-order expansion than the Lorenz equations. The variables $\bar{\psi}$ and $\bar{\theta}$ are expanded in double Fourier series in $\bar{x}$ and $\bar{y}$. The coefficients $\bar{\psi}_{m,n}$ and $\bar{\theta}_{m,n}$ represent the terms $\sin(2\pi m\bar{x}/\bar{\lambda})$ or $\cos(2\pi m\bar{x}/\bar{\lambda})$ and $\sin(\pi n\bar{y})$ in the expansions. In the limit of infinite Prandtl number the left-hand side of (12.14) is zero resulting in a linear relation between $\bar{\psi}_{m,n}$ and $\bar{\theta}_{m,n}$ that can be written

$$\bar{\psi}_{m,n} = -\left(\frac{2\pi m}{\bar{\lambda}}\right)\frac{\bar{\theta}_{m,n}}{(4\pi^2 m^2/\bar{\lambda}^2 + \pi^2 n^2)^2} \tag{13.4}$$

This result is then substituted into (12.15). The lowest consistent order of truncation beyond that used by Lorenz is $m = 2$ for the expansion in x ($m = 0, 1, 2$) and $n = 4$ for the expansion in y ($n = 1, 2, 3, 4$). This truncation yields a set of 12 ordinary differential equations for the time dependence of the temperature coefficients $\bar{\theta}_{m,n}$ that can be written

$$\frac{d\bar{\theta}_{m,n}}{dt} = -\sum_{p=-2}^{2}\sum_{q=-4}^{4}\frac{2\pi^2}{\bar{\lambda}}(mq - np)\left(\frac{2\pi p}{\bar{\lambda}}\right)\frac{\bar{\theta}_{p,q}\bar{\theta}_{m-p,n-q}}{[(4\pi^2 p^2/\bar{\lambda}^2) + \pi^2 q^2]^2}$$
$$+ Ra\left(\frac{2\pi m}{\bar{\lambda}}\right)^2\frac{\bar{\theta}_{m,n}}{[(4\pi^2 m^2/\bar{\lambda}^2) + \pi^2 n^2]^2} - \pi^2\left(\frac{4m^2}{\bar{\lambda}^2} + n^2\right)\bar{\theta}_{m,n} \tag{13.5}$$

It is necessary to take the resolution in the vertical direction twice compared with the horizontal direction in order to resolve the convection terms in the energy equation.

The time evolution of the 12 coefficients $\bar{\theta}_{0,1}, \bar{\theta}_{0,2}, \bar{\theta}_{0,3}, \bar{\theta}_{0,4}, \bar{\theta}_{1,1}, \bar{\theta}_{1,2}, \bar{\theta}_{1,3}, \bar{\theta}_{1,4}, \bar{\theta}_{2,1}, \bar{\theta}_{2,2}, \bar{\theta}_{2,3}, \bar{\theta}_{2,4}$ is found by integrating numerically the 12 equations given by (13.5). The time evolution can be thought of as trajectories in a 12-dimensional phase space. It is convenient to project the 12-dimensional trajectories onto the two-dimensional phase space consisting of $\bar{\theta}_{1,1}$ and $\bar{\theta}_{2,1}$; these correspond to the fundamental mode and the first subharmonic. There are two parameters in this problem, the Rayleigh number, Ra or r, and the wavelength. In this discussion solutions are given only for the critical value of the wavelength $\bar{\lambda} = 2^{3/2}$.

At the subcritical Rayleigh numbers $0 < r < 1$ ($0 < Ra < 657.512$) the only fixed point of the solution is at the origin and it is stable; there is no flow. For higher Rayleigh numbers, the two fixed points corresponding to clockwise and counterclockwise rotations in the fundamental model $\bar{\theta}_{1,1}$ become stable. The steady-state solution for $Ra = 10^4$ ($r = 15.21$) is given in Table 13.1. It is seen that only six of the 12 coefficients are nonzero: $\bar{\theta}_{0,2}, \bar{\theta}_{0,4}, \bar{\theta}_{1,1}, \bar{\theta}_{1,3}, \bar{\theta}_{2,2}$, and $\bar{\theta}_{2,4}$. This solution was obtained by specifying an initial condition near the origin and studying the time evolution of the 12 coefficients using (13.5). This time evolution projected onto the $\bar{\theta}_{2,1}\bar{\theta}_{1,1}$-plane is given

Table 13.1. *Numerical values of Fourier coefficients of the fixed points of the 12-mode equations* (13.5)

	Ra		
	10^4	3×10^4	4.3×10^4
$r = Ra/R_c$	15.21	45.62	65.39
$\bar{\theta}_{0,1}$	0.000	0.000	−2120.222
$\bar{\theta}_{0,2}$	1506.097	4571.719	6663.231
$\bar{\theta}_{0,3}$	0.000	0.000	−366.705
$\bar{\theta}_{0,4}$	286.133	2191.940	3355.777
$\bar{\theta}_{1,1}$	506.978	0.000	0.000
$\bar{\theta}_{1,2}$	0.000	0.000	0.000
$\bar{\theta}_{1,3}$	66.170	0.000	0.000
$\bar{\theta}_{1,4}$	0.000	0.000	0.000
$\bar{\theta}_{2,1}$	0.000	1117.832	−1415.752
$\bar{\theta}_{2,2}$	426.445	0.000	−398.378
$\bar{\theta}_{2,3}$	0.000	509.313	−623.561
$\bar{\theta}_{2,4}$	105.531	0.000	−490.614

in Figure 13.1(a). Although the subharmonic coefficient $\bar{\theta}_{2,1}$ is zero at the fixed point, it is nonzero during the time evolution.

The steady-state solution for $Ra = 3 \times 10^4$ ($r = 45.62$) is also given in Table 13.1. It is seen that only four of the 12 coefficients are nonzero: $\bar{\theta}_{0,2}, \bar{\theta}_{0,4}, \bar{\theta}_{2,1}, \bar{\theta}_{2,3}$. At this Rayleigh number the fundamental mode and its associated coefficients $\bar{\theta}_{1,1}, \bar{\theta}_{1,3}, \bar{\theta}_{2,2}$, and $\bar{\theta}_{2,4}$ are zero at the stable fixed point. The time evolution of the solution projected onto the $\bar{\theta}_{2,1}\bar{\theta}_{1,1}$-plane is given in Figure 13.1(b). Finally, the steady-state solution for $Ra = 4.3 \times 10^4$ ($r = 65.39$) is given in Table 13.1. It is seen that eight of the 12 coefficients are now nonzero: $\bar{\theta}_{0,1}, \bar{\theta}_{0,2}, \bar{\theta}_{0,3}, \bar{\theta}_{0,4}, \bar{\theta}_{2,1}, \bar{\theta}_{2,2}, \bar{\theta}_{2,3}, \bar{\theta}_{2,4}$. All of the $\bar{\theta}_{1,n}$ coefficients are zero including the fundamental mode $\bar{\theta}_{1,1}$. The time evolution of the solution projected onto the $\bar{\theta}_{2,1}\bar{\theta}_{1,1}$-plane is given in Figure 13.1(c). It is seen that the evolution prior to entering the stable fixed point is much more complex; the solution oscillates in the positive $\bar{\theta}_{2,1}$ quadrants before entering the negative quadrants.

The time evolution of the solution for $Ra = 4.5 \times 10^4$ ($r = 68.44$) is given in Figure 13.1(d); it is fully chaotic and no fixed points are

Figure 13.1. Numerical solutions of the 12-mode infinite Prandtl number equations projected onto the $\bar{\theta}_{2,1}\bar{\theta}_{1,1}$-plane of the 12-dimensional phase space.
(a) $Ra = 10\,000\,(r = 15.21)$,
(b) $Ra = 30\,000\,(r = 45.62)$,
(c) $Ra = 43\,000\,(r = 65.39)$,
(d) $Ra = 45\,000\,(r = 65.39)$.

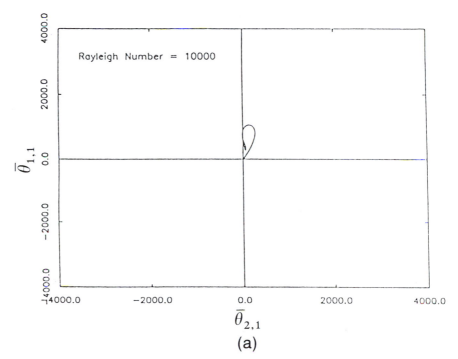

(a)

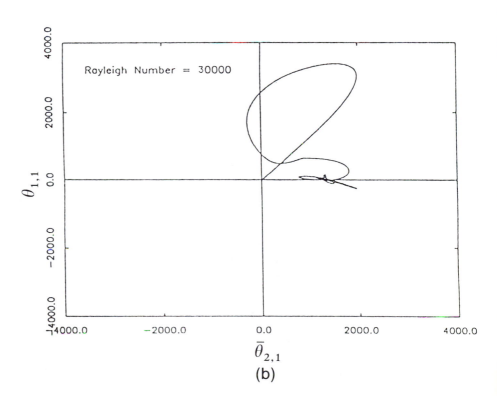

(b)

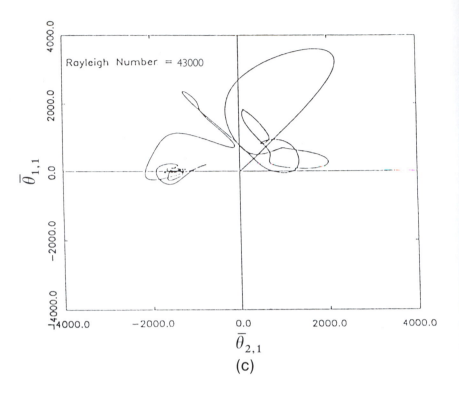

(c)

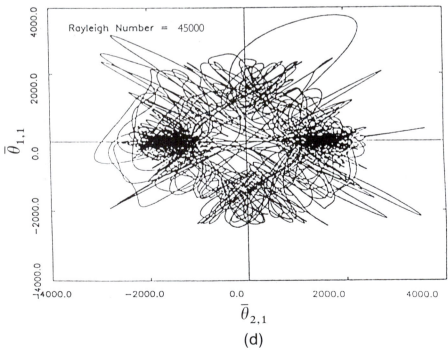

(d)

stable. The flow alternates between aperiodic oscillations about the two fundamental modes (clockwise and counterclockwise) and the two subharmonic modes (clockwise and counterclockwise). All 12 coefficients are nonzero and are time dependent. The time dependences of the $\bar{\theta}_{2,1}$ and $\bar{\theta}_{1,1}$ coefficients are shown in Figure 13.2. The resemblance to the time behavior of the Lorenz attactror illustrated in Figure 12.2(c) is striking. Oscillatory behavior of the $\bar{\theta}_{2,1}$ mode amplifies until the flow undergoes a turbulent burst in the fundamental $\bar{\theta}_{1,1}$ mode, where it is briefly trapped before flipping into the $\bar{\theta}_{2,1}$ mode with the opposite sense of rotation.

In order to better understand the transitions in the behavior of the time-dependent solutions Figure 13.3 gives two projections of the loci of fixed points as a function of the Rayleigh number of the system. The solid lines denote stable fixed points, the dashed lines denote unstable fixed points, and the open circles denote Hopf bifurcations. Figure 13.3(a) shows a projection onto the $Ra, \bar{\theta}_{1,1}$-plane. The origin representing the conduction solution becomes unstable and bifurcates at $Ra = 657.512$, giving two stable symmetric solutions that do not contain the $\bar{\theta}_{2,1}$ mode. One branch of this solution appears in the positive quadrant and is labeled $\bar{\theta}_{1,1}$(pure) in Figure 13.3(a) to distinguish it from the mixed-mode solution, which contains a contribution from the $\bar{\theta}_{2,1}$ mode. Each branch becomes unstable and undergoes a subcritical pitchfork bifurcation at $Ra = 3.802 \times 10^4$, producing four unstable mixed-mode solutions, labeled '$\bar{\theta}_{1,1}$(mixed)'. Each $\bar{\theta}_{1,1}$ mixed-mode branch sweeps back to a saddle bifurcation at $Ra = 1.909 \times 10^4$. This type of bifurcation configuration (subcritical pitchfork plus two saddles) typically produces hysteresis effects when the saddle has one stable branch and one unstable branch. Here, the $\bar{\theta}_{1,1}$(mixed) solution has one stable manifold (out of 12) on one side of the saddle, and two unstable manifolds on the other.

The second bifurcation of the conduction solution is at $Ra = 1315.023$, where two unstable symmetric fixed points dominant in the subharmonic $\bar{\theta}_{2,1}$ mode appear. Since these fixed points contain no component in the $\bar{\theta}_{1,1}$ mode, we call these '$\bar{\theta}_{2,1}$(pure)' unstable solution branches. Each of these branches becomes stable and undergoes a pitchfork bifurcation at $Ra = 2041.918$, resulting in the branching solution labeled '$\bar{\theta}_{2,1}$(mixed)' in Figures 13.3(a) and 13.3(b).

Figure 13.2. Time
dependences of the
coefficients (a) $\bar{\theta}_{1,1}$ and (b)
$\bar{\theta}_{2,1}$ for the solution given
in Figure 13.1(d);
$Ra = 45\,000$.

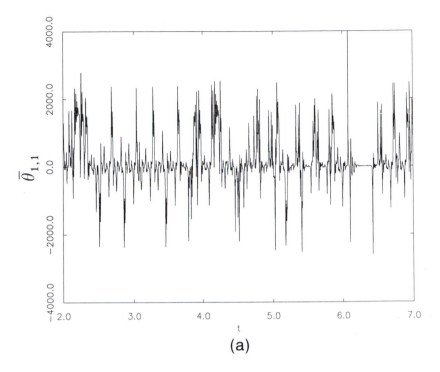

(a)

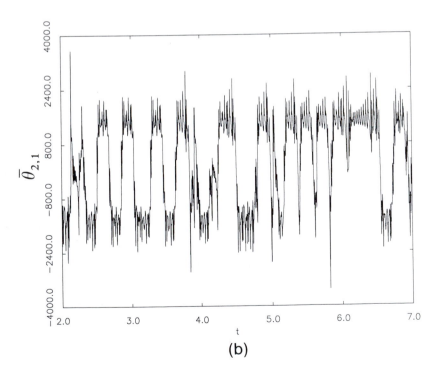

(b)

Figure 13.3. Bifurcation diagram for the 12-mode infinite Prandtl number equations. The fixed points of (13.5) are projected (a) onto the $Ra\bar{\theta}_{1,1}$-plane, (b) onto the $Ra\bar{\theta}_{2,1}$-plane. Stable branches are shown as solid lines, unstable branches as dashed lines, pitchfork bifurcations as solid circles, and Hopf bifurcations as open circles.

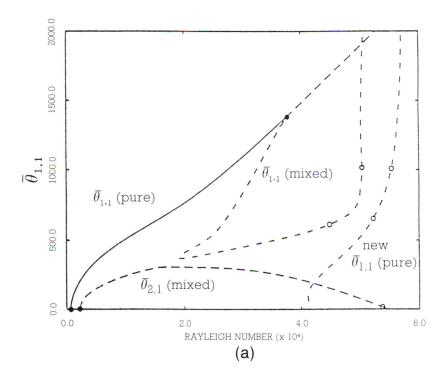

(a)

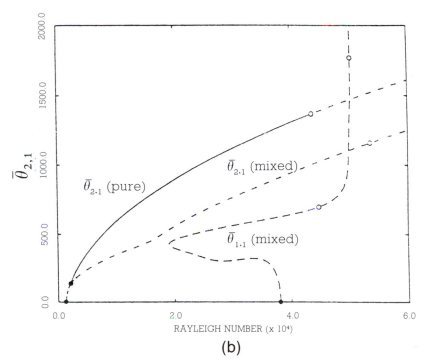

(b)

The $\bar{\theta}_{2,1}$ mixed-mode branches nearly connect with the $\bar{\theta}_{1,1}$ mixed-mode branches.

Both the fundamental and the subharmonic pure-mode solutions are stable between $Ra = 2369$ and $Ra = 3.802 \times 10^4$. The trajectories in Figures 13.1(a) and 13.1(b) have the same initial condition, yet the trajectory in Figure 13.1(a) converges to the fundamental subharmonic stable fixed point. Presumably this is because the unstable mixed-mode branches disrupt the separatrix between the basin of the attraction of the fundamental and subharmonic pure-mode solutions. Note that the transition from Figure 13.1(a) to Figure 13.1(b) occurs at a Rayleigh number above the stability limit of the fundamental mode.

The third bifurcation of the conduction solution is at $Ra = 4.140 \times 10^4$, where two unstable symmetric solutions in the fundamental mode $\bar{\theta}_{1,1}$ appear. These are labeled 'new $\bar{\theta}_{1,1}$(pure)' in Figure 13.3(a). Each of these undergo Hopf bifurcations at $Ra = 5.23 \times 10^4$ and $Ra = 5.53 \times 10^4$. At no point does the origin itelf undergo a Hopf bifurcation, nor does the conduction solution bifurcate to mixed-mode solutions.

We detected no Hopf bifurcations for the conduction or fundamental harmonic solutions; however, the stable subharmonic branch undergoes two Hopf bifurcations (Figure 13.3(b)), one at $Ra = 4.37 \times 10^4$ and one at $Ra = 6.36 \times 10^4$. Each mixed-mode saddle ($\bar{\theta}_{1,1}$(mixed)) undergoes two Hopf bifurcations, at $Ra = 4.491 \times 10^4$ and at $Ra = 5.039 \times 10^4$. Each mixed-mode saddle ($\bar{\theta}_{2,1}$(mixed)) undergoes six Hopf bifurcations. Each of these Hopf bifurcations sheds stable or unstable periodic orbits that are responsible for the oscillations of the trajectory at $Ra = 4.5 \times 10^4$ shown in Figure 13.1(d). In Figure 13.4, the first 7000 points of the trajectory at $Ra = 4.5 \times 10^4$ are projected onto the $\bar{\theta}_{2,1}\bar{\theta}_{1,1}$-plane (dotted line) and are shown superimposed on the central portion of the branches of the fixed points (solid lines). Note that the trajectory weaves aperiodically around several Hopf bifurcations (open circles).

Physically, infinite Prandtl number, high Rayleigh number convection becomes time dependent through boundary layer instabilities that generate thermal plumes. In terms of spectral expansions, these instabilities result from the nonlinear coupling in the convective terms of the heat equation.

Figure 13.4. Loci of the fixed points from Figure 13.3 projected onto the $\bar{\theta}_{2,1}\bar{\theta}_{1,1}$-plane. Superimposed as a dotted line is the time evolution of the chaotic solution from Figure 13.1d.

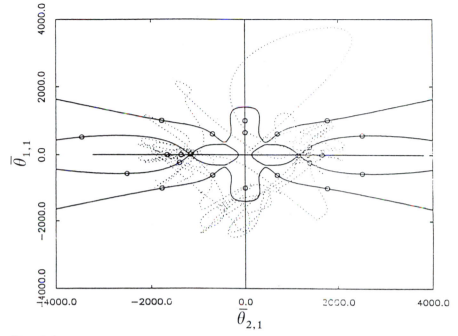

Problems

Problem 13.1. Show that the temperature coefficient B in the Lorenz expansion (12.24) is related to θ_{11} by

$$B = -\frac{8^{1/2}\bar{\lambda}^2\theta_{11}}{\pi^3(4+\bar{\lambda}^2)^3}$$

For $Ra = 10^4$ and $\bar{\lambda} = 8^{1/2}$ compare the value of B from the three-mode (Lorenz) expansion with the value from the 12-mode expansion.

Problem 13.2. Show that the temperature coefficient C in the Lorenz expansion (12.24) is related to θ_{02} by

$$C = \frac{4\bar{\lambda}^4\theta_{02}}{\pi^3(4+\bar{\lambda}^2)^3}$$

For $Ra = 10^4$ and $\bar{\lambda} = 8^{1/2}$ compare the value of C from the three-mode (Lorenz) expansion with the value from the 12-mode expansion.

CHAPTER FOURTEEN

Rikitake dynamo

The earth's magnetic field is attributed to the electrically conducting outer core, which acts as a dynamo. The liquid outer core is primarily composed of iron which is an excellent electrical conductor at core conditions. Electrical currents in the core generate a magnetic field. Buoyancy forces in the core, due to either temperature or composition, drive a fluid flow. The flowing electrical conductor in the magnetic field induces an electric field. This is a self-excited dynamo.

Paleomagnetism is the study of the earth's past magnetic field from the records preserved in magnetized rocks. Rocks containing small amounts of ferromagnetic minerals such as magnetite and hematite can acquire a weak permanent magnetism when they are formed. This fossil magnetism in a rock is referred to as natural remanent magnetism.

Many volcanic rocks at the surface of the earth can be magnetized because of the presence of minerals such as magnetite. When these volcanic rocks were cooled through the Curie temperature they acquired a permanent magnetism from the earth's field at the time of cooling. Paleomagnetic studies of remanent magnetism have provided a variety of remarkable conclusions. These studies have traced the movement of the rocks due to plate tectonics and continental drift over periods of hundreds of millions of years. They have shown that the magnetic field at the surface of the earth has been primarily a dipole, as it is today, and has remained nearly aligned to the earth's axis of rotation. These studies have also shown that the earth's magnetic field has been subject to random reversals in which the north magnetic pole becomes the south magnetic pole and *vice versa*. The observed polarities of the earth's magnetic field for the last 10 million years (Myrs) are given in Figure 14.1. Measurements indicate that for the last 720 000 years the magnetic field has been in its present (normal) orientation; between 0.72 and 2.5 Myr ago there was a period during which the orientation of the field was predominantly reversed. Clearly one characteristic of the core dynamo is that it is subject to spontaneous reversals.

No detailed theory exists for the behavior of the core dynamo. The viscosity of the liquid outer core is sufficiently small that the flow is undoubtedly turbulent. Thus the patterns of flow, electrical currents, and magnetic fields are very complex. Because of this complexity relatively simple disk dynamos have been proposed as analog models. Rikitake (1958) proposed the symmetrical two-disk dynamo illustrated in Figure 14.2. It is composed of two symmetrical disk dynamos in which the current produced by one dynamo energizes the other. Equal torques G are applied to the two dynamos in order to overcome ohmic losses. Rikitake (1958) found that this dynamo was subject to random reversals of the magnetic field but it was much later (Cook and Roberts, 1970) that it was demonstrated that the Rikitake dynamo behaved in a chaotic manner.

The steady-state behavior of the Rikitake dynamo is relatively easy to understand as illustrated in Figure 14.2. The current I_1 passes through the current loop on the right in a clockwise direction. This current loop generates a magnetic field B_2 that is positive in the downward direction. This magnetic field also passes through the rotating, electrically conducting disk on the right. Because of the applied torque G the disk is rotating in the counterclockwise direction with an angular velocity Ω_2. The induced electric field in the disk $\mathbf{E}_i = \mathbf{u} \times \mathbf{B}$ is the inward radial direction as illustrated. This induced electrical field drives the electric current I_2. In the steady state the applicable circuit equation is

$$RI_2 = M\Omega_2 I_1 \tag{14.1}$$

where R is the resistance in either circuit and M is the mutual inductance between the current loops and the electrically conducting disks.

An electrical current $\mathbf{I}$ in a magnetic field $\mathbf{B}$ results in the

Figure 14.1. Observed polarity of the earth's magnetic field for the last 10 Myr; the scale is in Myr before the present (BP). The solid bands are the normal (present) polarity and the open bands are reversed polarity. The last polarity reversal occurred 720 000 years ago.

electromotive force $\mathbf{F} = \mathbf{I} \times \mathbf{B}$ per unit length of current path. The interaction between the magnetic field B_2 and the radially inward electric current I_2 results in a torque in the clockwise direction. In the steady state this torque balances the applied torque G and is given by

$$G = MI_1I_2 \tag{14.2}$$

The discussion given above is also applicable to the current loop and rotating disk on the left in Figure 14.2.

The equations governing the unsteady behavior of the two-disk Rikitake dynamo are

$$L\frac{dI_1}{dt} + RI_1 = M\Omega_1 I_2 \tag{14.3}$$

$$L\frac{dI_2}{dt} + RI_2 = M\Omega_2 I_1 \tag{14.4}$$

$$C\frac{d\Omega_1}{dt} = G - MI_1I_2 \tag{14.5}$$

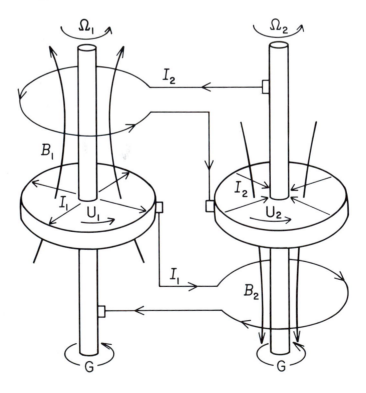

Figure 14.2. Illustration of the Rikitake two-disk dynamo. The applied torques G drive the shafts and electrically conducting disks to rotate in the counterclockwise direction. The currents in the coils generate magnetic fields that interact with the rotating disks to generate the currents.

$$C\frac{d\Omega_2}{dt} = G - MI_1I_2 \tag{14.6}$$

where L is the self-inductance in either circuit and C the moment of inertia of each dynamo. Subtracting (14.5) and (14.6) and integrating gives

$$\Omega_1 - \Omega_2 = \Omega_0 \tag{14.7}$$

where Ω_0 is a constant. This can replace either (14.5) or (14.6).

Again it is appropriate to introduce nondimensional variables and parameters according to

$$\tau = \left(\frac{GM}{LC}\right)^{1/2} t, \; X_1 = \left(\frac{M}{G}\right)^{1/2} I_1, \; X_2 = \left(\frac{M}{G}\right)^{1/2} I_2, \; Y_1 = \left(\frac{CM}{GL}\right)^{1/2} \Omega_1,$$

$$Y_2 = \left(\frac{CM}{GL}\right)^{1/2} \Omega_2, \; A = \left(\frac{CM}{GL}\right)^{1/2} \Omega_0, \; \mu = \left(\frac{CR^2}{GLM}\right)^{1/2} \tag{14.8}$$

Substitution into (14.3)–(14.5) and (14.7) gives

$$\frac{dX_1}{d\tau} + \mu X_1 = Y_1 X_2 \tag{14.9}$$

$$\frac{dX_2}{d\tau} + \mu X_2 = (Y_1 - A)X_1 \tag{14.10}$$

$$\frac{dY_1}{d\tau} = 1 - X_1 X_2 \tag{14.11}$$

$$Y_2 = Y_1 - A \tag{14.12}$$

This is a set of three coupled, nonlinear differential equations that determine the time evolution of the Rikitake dynamo. Setting the time derivatives equal to zero the steady-state solutions are obtained:

$$X_1 = \pm K \tag{14.13}$$
$$X_2 = \pm K^{-1} \tag{14.14}$$
$$Y_1 = \mu K^2 \tag{14.15}$$
$$Y_2 = \mu K^{-2} \tag{14.16}$$

where

$$A = \mu(K^2 - K^{-2}) \tag{14.17}$$

The plus and minus signs refer respectively to the normal and reversed states of the magnetic field.

Stability calculations (Cook and Roberts, 1970) have shown that the singular points given above are unstable for all parameter values.

Their numerical solutions for $\mu = 1$ and $K = 2$ are given in Figures 14.3 and 14.4. The singular points $X_2 = \pm\frac{1}{2}$, $Y_1 = 4$ are shown in the X_2Y_1 phase plane illustrated in Figure 14.3. The strange-attractor behavior of the solution is very similar to that of the Lorenz equations given in Figures 12.1(a), (b). The time evolution of the solutions, given in Figure 14.4, is also similar to that of the Lorenz equations given in Figure 12.1(c). Oscillations grow in one polarity of the field until it flips into the other polarity. There are also strong similarities between the reversals of polarity for the Rikitake dynamo and the reversals in polarity of the earth's magnetic field illustrated in Figure 14.1.

The Rikitake model is clearly a gross simplification of the complex fluid flows that occur in the earth's core. Nevertheless, the model produces a pattern of random reversals that is remarkably similar to the reversals of the earth's magnetic field. Again this can be taken as evidence that the dynamo action in the core is chaotic. It should be emphasized, however, that there is a fundamental difference between the Lorenz equations and the Rikitake dynamo equations. The Lorenz equations are derived directly from the appropriate equations for thermal convection. The Rikitake dynamo equations

Figure 14.3. Numerical solution of the Rikitake two-disk dynamo equations (14.9)–(14.11) in the $X_1X_2Y_1$ phase space projected onto the X_2Y_1 phase plane. The singular points corresponding to normal and reversed polarizations of the magnetic field, $X_2 = \pm 0.5$, $Y_1 = 4$ are shown.

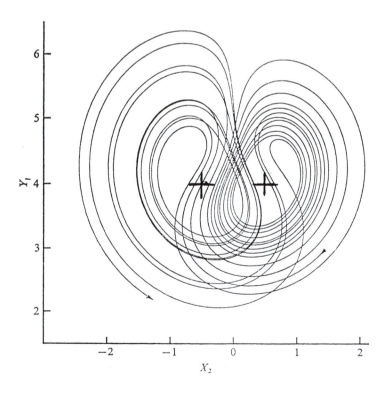

are rigorously applicable to the model dynamo but the relation of the model dynamo to the dynamics of the core is completely *ad hoc*.

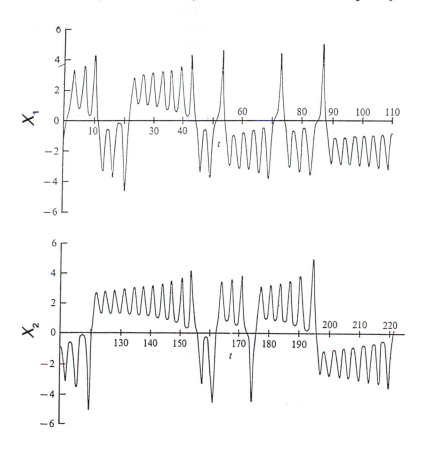

Figure 14.4. Time evolution of X_1 and X_2 for the solution given in Figure 14.3.

Problem

Problem 14.1. Consider a single disk dynamo. Determine the steady state current I and angular velocity Ω in terms of resistance R, mutual inductance M, and applied torque G.

CHAPTER FIFTEEN

Renormalization group method

In the first eight chapters of this book we considered the fractal behavior of natural systems. This behavior was generally statistical and the physical causes were generally inaccessible. In the six chapters that followed we considered low-order dynamical systems that exhibit chaotic behavior. Because of the low order, the examples are generally quite far removed from natural systems of interest. In this chapter and the next we take a collective view of natural phenomena and consider some applications in geology and geophysics.

Thermodynamics represents the standard approach to collective phenomena. System variables are defined, i.e. temperature, pressure, density, entropy; and the evolution of these variables is determined from the first law of thermodynamics (conservation of energy) and the second law of thermodynamics (variation of entropy). Statistical mechanics provides the rational microscopic basis for much of thermodynamics.

In general, neither thermodynamics nor statistical mechanics yields fractal statistics or chaotic behavior. Exceptions include critical points and phase changes. A characteristic feature of a phase change is a discontinuous (catastrophic) change of macroscopic parameters of the system under a continuous change in the system's state variables. For example, when water freezes its viscosity changes from a very small value to a very large value with no change in temperature.

The renormalization group method has been used successfully in treating a variety of phase change and critical-point problems (Wilson and Kagut, 1974). This method often produces fractal statistics and explicitly utilizes scale invariance. A relatively simple system is considered at the smallest scale; the problem is then renormalized (rescaled) in order to utilize the same system at the next larger scale. The process is repeated at larger and larger scales. This is very similar to our renormalization models for fragmentation and the concentration of ores given in Chapters 3 and 5 respectively.

We consider three specific applications in order to illustrate the use of the renormalization group method. We first consider a model for the flow of a fluid through a porous-medium. Such problems have wide applicability in ground-water hydrology and petroleum engineering. The porous media may either be a granular material such as a sandstone or a matrix of rock fractures. In either case Darcy's law is generally applicable and the fluid velocity is proportional to the pressure gradient; Darcy's law is a linear relation.

We apply the renormalization group method in order to understand the onset of fluid flow through a porous medium. We first consider the two-dimensional model illustrated in Figure 15.1(a). A square array is made up of a matrix of square boxes that are referred to as elements; each box (element) may be either permeable or impermeable. For the example given in Figure 15.1(a) there are $n = 256$ elements. The probability that an element is permeable is p_0, the probability it is impermeable is $1 - p_0$. The question is whether the square array is permeable or impermeable. The array is defined to be permeable if there is a continuous permeable path from the left side of the array to the right side of the array. This is clearly a statistical question since the actual distribution of permeable and impermeable elements is random. For a specified value of p_0 there is a probability P that the array with n elements is permeable. For large arrays it is found that P is very small ($P \ll 1$) if $0 < p_0 < p^*$ where p^* is the critical

Figure 15.1. (a) A 16×16 array of square elements. The probability p_0 that an element is permeable is 0.5; either the dark or the light elements can be assumed to be permeable. For either case, no permeable path across the array is found. (b) Illustration of the renormalization group method; four square elements are considered at each of the four scales.

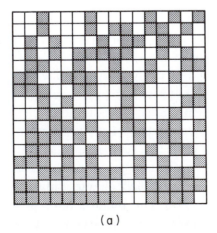

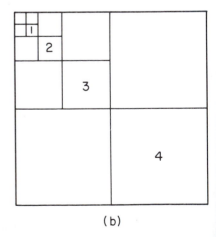

(a)   (b)

probability for the percolation threshold and P is near unity ($P \approx 1$) if $p^* < p_0 < 1$ (Stinchcombe and Watson, 1976). Thus there is a critical value of p_0, p^*, for the onset of flow through the grid of elements. If p_0 is less than this critical value p^* a large square grid will almost certainly be impermeable to flow. If p_0 is greater than this critical value p^* a large square grid will almost certainly be permeable to flow.

It may be easier to visualize this problem if one considers a forest made up of a square grid of trees. The probability that a grid point has a tree is p_0. The question is whether a forest fire can burn through the forest if a tree can ignite only its nearest neighbors. If there are no nearest neighbors the fire does not spread. This forest fire problem is mathematically identical to the percolation problem considered above.

We now turn to the percolation problem and consider in some detail the 16×16 array of square elements illustrated in Figure 15.1(a). The total number of elements $n = 256$. Taking $p_0 = 0.5$, it has been randomly determined whether each element is permeable or impermeable. With $p_0 = 0.5$ either the dark squares or the light squares can be taken to be permeable. In either case no continuous permeable path is found either horizontally or vertically. Using the Monte Carlo approach a large number of random realizations would be carried out and the probability $P(p_0)$ that the array is permeable would be determined. For the two-dimensional square array with large n, numerical simulations find that the critical probability for the onset of flow in the array is $p^* = 0.59275$ (Stauffer, 1985).

It is also of interest to consider the number–size statistics for percolation clusters at the critical limit $p_0 = p^*$. The size of a percolation cluster is defined to be the number of permeable elements in contact with each other when the array first becomes permeable. The number of elements in the percolation cluster n_e^* has been determined numerically as a function of the array size n. For two-dimensional arrays it is found that (Stauffer, 1985)

$$n_e^* \sim n^{91/96} \tag{15.1}$$

This result can be compared with the deterministic Sierpinski carpet assuming that the remaining squares illustrated in Figure 2.2 represent a percolation cluster. For the Sierpinski carpet $n_e = 8$ when $n = 9$ and $n_e = 64$ when $n = 81$, thus

$$n_e = n^{\ln 8/2 \ln 3} = n^{D/2} \tag{15.2}$$

Comparison of (15.1) and (15.2) shows that the fractal dimension for the percolation cluster at the critical limit is $D = 91/48 = 1.896$ which is very close to the value for a Sierpinski carpet $D = \ln 8/\ln 3 = 1.893$.

The direct statistical approach to the percolation problem becomes extremely time consuming if the number of elements is large since many random realizations must be considered. An alternative approximate approach is illustrated in Figure 15.1(b). At the lowest order a square array of four elements is considered. The probability p_1 that the first-order cell is permeable, is determined in terms of the probability p_0 that an individual first-order element is permeable. The cell is defined to be permeable if there is a continuous permeable path from left to right. The problem is then renormalized and four first-order cells become the four second-order elements in a second-order cell. The probability p_2 that the second-order cell is permeable, is then determined in terms of the probability p_1 that a second-order element (first-order cell) is permeable. The process is repeated at larger and larger scales (higher and higher orders). This is the renormalization group method and is essentially the same as the fractal fragmentation model illustrated in Figure 3.2.

We now consider the first-order cell and determine the probability that it is permeable. All possible configurations are illustrated in Figure 15.2. The probability that all four elements are impermeable is $(1 - p_0)^4$ and there is only one configuration, as shown in Figure 15.2(a). This configuration is clearly not permeable $(-)$. The probability that one element is permeable and three elements are not is $p_0(1 - p_0)^3$ and there are four configurations for the cell as illustrated in Figure 15.2(b). The permeable element can be in any of the four positions shown but the cell is not permeable in any of them $(-)$.

Figure 15.2. A cell consisting of four elements is considered. A cell is defined to be permeable if there is a continuous permeable path from left to right. In (a) all four elements are impermeable, in (b) one element is permeable and the four possible configurations are given, in (c) two elements are permeable and the six possible configurations are given, in (d) three elements are permeable and the four possible configurations are given, and in (e) all four elements are permeable. The configurations that are permeable from left to right are denoted by $(+)$ and the configurations that are impermeable from left to right are denoted by $(-)$.

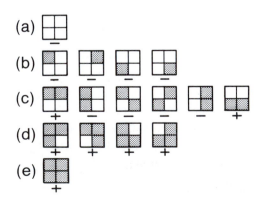

The probability that two elements are permeable and two elements are not is $p_0^2(1 - p_0)^2$ and there are six independent configurations as shown in Figure 15.2(c). The first and sixth configurations result in the cell being permeable for flows from left to right $(+)$ and the other four configurations are impermeable $(-)$. The probability that three elements are permeable and one element is not is $p_0^3(1 - p_0)$ and there are four independent configurations as shown in Figure 15.2(d). All four cell configurations are permeable $(+)$. The probability that all four elements are permeable is p_0^4 and there is only one configuration as shown in Figure 15.2(e): it is permeable $(+)$.

Taking into account all possible configurations the probability that the first-order cell is permeable is given by

$$p_1 = 2p_0^2(1 - p_0)^2 + 4p_0^3(1 - p_0) + p_0^4 = 2p_0^2 - p_0^4 \tag{15.3}$$

The first-order probability includes the two configurations with two permeable elements, the four configurations with three permeable elements, and the single configuration with four permeable elements. Renormalization is carried out and four first-order cells become second-order elements. After renormalization exactly the same statistics are applicable to the second-order cell with the result

$$p_2 = 2p_1^2 - p_1^4 \tag{15.4}$$

This result can be applied to the nth-order cell with the result

$$p_{n+1} = 2p_n^2 - p_n^4 \tag{15.5}$$

This recursive relation for the probability is quite similar to the logistic map considered in Chapter 10.

Figure 15.3 shows the dependence of p_{n+1} on p_n given in (15.5). In order to consider the fixed points it is appropriate to rewrite (15.5) as

$$f(x) = 2x^2 - x^4 \tag{15.6}$$

The fixed points are obtained by setting $f(x) = x$ with the result

$$x = 2x^2 - x^4 \tag{15.7}$$

In the range $0 < x < 1$ there are three fixed points $x = 0, 0.618, 1$. The corresponding values of $\lambda = df/dx$ are 0, 1.528, 0. The fixed points at $x = 0$ and 1 are stable since $|\lambda| < 1$ but the fixed point at $x = 0.618$ is unstable.

In order to illustrate further the iteration of the probabilities given by (15.5) we consider two specific cases. For $p_0 = 0.4$ we find $p_1 = 0.294$, $p_2 = 0.166$, and $p_3 = 0.054$ as illustrated in Figure 15.3. The construction is the same as that used in Figures 10.1–10.5. As the iteration is continued to large n the probability p_n approaches the stable fixed point $p_n = 0$. A large two-dimensional array is impermeable for $p_0 = 0.4$. For $p_0 = 0.8$ we find $p_1 = 0.870, p_2 = 0.941$, and $p_3 = 0.987$ as illustrated in Figure 15.3. As the iteration is continued to large n the probability p_n approaches the stable fixed point $p_n = 1$. A large two-dimensional array is permeable for $p_0 = 0.8$.

The unstable fixed point at $p^* = 0.618$ is a critical point. At the critical point, $p_n = p^*$ for all values of n; the probability that a cell is permeable is scale invariant. For probabilities smaller than the critical value $0 < p_0 < p^*$ the iteration is to the impermeable limit $p_n = 0$. For probabilities greater than the critical value $p^* < p_0 < 1$ the iteration is to the permeable limit $p_n = 1$. The value $p^* = 0.618$ obtained by the renormalization group method compares with $p^* = 0.59275$ obtained by the direct numerical simulations for large arrays.

The two-dimensional renormalization group method can be based on a larger lowest-order array. Consider a 3×3 array with nine elements. Taking into acount all possible configurations the probability p_{n+1} that the $(n + 1)$th-order cell is permeable is related to the probability p_n that the nth-order cell is permeable by

$$p_{n+1} = 3p_n^3 + 3p_n^4 - 2p_n^5 - 15p_n^6 + 18p_n^7 - 7p_n^8 + p_n^9 \qquad (15.8)$$

The behavior of this relation is essentially similar to the behavior of

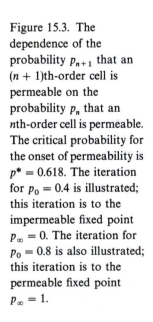

Figure 15.3. The dependence of the probability p_{n+1} that an $(n + 1)$th-order cell is permeable on the probability p_n that an nth-order cell is permeable. The critical probability for the onset of permeability is $p^* = 0.618$. The iteration for $p_0 = 0.4$ is illustrated; this iteration is to the impermeable fixed point $p_\infty = 0$. The iteration for $p_0 = 0.8$ is also illustrated; this iteration is to the permeable fixed point $p_\infty = 1$.

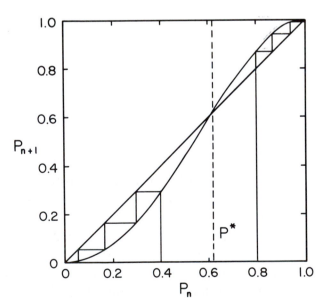

(15.5). The critical value for the onset of permeability is $p^* = 0.609$. This is somewhat closer to the numerical value $p^* = 0.59275$ than the 2×2 renormalization group result given above.

Extensive studies have also been carried out on three-dimensional cubic arrays made up of n cubic elements. In this case the individual cubic elements are taken to be either permeable or impermeable. The array is defined to be permeable if there is a continuous permeable path from one side of the array to the other. Numerical studies of large cubic arrays show that the critical probability for the onset of flow in the array is $p^* = 0.3117$ (Stauffer, 1985). Again, a fractal relation of the form (15.2) is obtained between the number of permeable elements n_e in the critical percolation cluster and the total number of elements n with $D = 2.5$. This result can be compared with the deterministic Menger sponge illustrated in Figure 2.3(a), assuming that the remaining cubes represent a percolation cluster. For the Menger sponge $n_e = 20$ when $n = 27$ and $n_e = 400$ when $n = 729$; the value $D = 2.727$ for the Menger sponge is somewhat higher than the value for three-dimensional percolation clusters.

The simplest renormalization group model for the array of cubic elements is a $2 \times 2 \times 2$ cubic array of eight elements. Taking into account all possible configurations the probability that the $(n + 1)$th-order cell is permeable is related to the probability that the nth-order cell is permeable by

$$p_{n+1} = 4p_n^2 - 6p_n^4 + 4p_n^6 - p_n^8 \qquad (15.9)$$

The critical value for the onset of permeability is $p^* = 0.282$. This is in reasonably good agreement with the numerical results considering the simplicity of the model.

For fluid flow through rocks the two measurable quantities are the porosity (the degree to which void space becomes filled with fluid) and the permeability (the ability of the fluid to flow through the rock under fluid pressure). The highly idealized model considered above predicts that there will be a sudden onset of permeability at a critical value of the porosity. Although rocks with low porosity have essentially zero permeability, the sudden onset of permeability at a universal critical value of the porosity is not observed. This is attributed to the variety of aperture sizes and lengths occurring in natural systems, which is not fully described by the idealized model.

Problems associated with electrical conduction through a matrix of elements are essentially identical to the percolation problems considered above. Madden (1983) has applied the renormalization group method to the onset of electrical conductance through a grid of electrical conductors and insulators.

Our second application of the renormalization group method is to fragmentation (Allègre *et al.*, 1982; Turcotte, 1986a). We again consider fragmentation in terms of the scale-invariant model illustrated in Figure 3.2 with fragments that differ in size by factors of two. However, there is a fundamental difference. In Chapter 3 we considered a large zero-order cell and made smaller and smaller elements. Here we consider the first-order elements to be the smallest fragments and make larger and larger cells following the renormalization group method. At the lowest order a cubic array of eight elements is assumed to constitute a first-order cell. Following Allègre *et al.* (1982), each element in a cell is hypothesized to be either fragile (f) if it is permeated with microfractures or sound (s) if it is not. The probability that a first-order cell is fragile is p_1 and is determined in terms of the probability that an individual element is fragile, p_0. It is necessary to specify a condition for the fragility (soundness) of the first-order cell. Allègre *et al.* (1982) hypothesized that a cell was sound if a 'pillar' of sound elements linked two faces of the cell. Having prescribed the probability p_0 that an element is fragile it is necessary to determine the probability p_1 that a cell is fragile. In order to do this it is necessary to consider all alternative configurations. In each cell there can be zero to eight fragile elements so that there are $2^8 = 256$ possible combinations. Excluding multiplicities, there are 22 topologically different configurations; these are illustrated in Figure 15.4. The numbers in parentheses are the multiplicities of each configuration. The fragile elements are indicated by solid dots at the corners. Configurations 4f, 5c, 6b, 6c, 7 and 8 are fragile and are indicated by solid underlining in Figure 15.4. Examples of sound and fragile cells are illustrated in Figure 15.5; 5b is a sound cell with the pillar of strength illustrated by heavy lines, and 5c is a fragile cell.

The probability that all eight elements are fragile is p_0^8, the probability that seven elements are fragile and one is sound is $p_0^7(1 - p_0)$, and so forth. Taking into account the configurations that are fragile and their multiplicities the probability p_1 that a first-order cell is fragile is related to the probability p_0 that a first-order element

Figure 15.4. Illustration of 22 topologically different configurations. The 'fragile' elements are indicated by solid dots on the corners. The first number beneath a configuration indicates the number of fragile elements, the letters indicate the various configurations with the same number of fragile elements, and the number in parentheses gives the multiplicity of that configuration. Using the Allègre *et al.* (1982) 'pillar of strength' condition, the fragile cells are underlined.

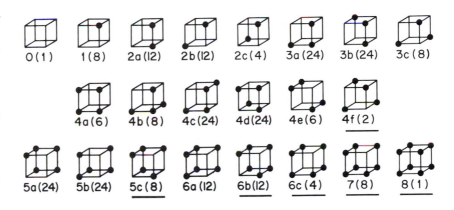

Figure 15.5. Eight cubic elements with dimension h form a cell with dimension $2h$. Configurations 5b and 5c are illustrated; the five fragile elements are shaded, and the three sound elements are unshaded. Cell 5b is sound because the unshaded (outlined) 'pillar' of sound elements links two faces. Cell 5c is fragile because no pillar of sound elements links two faces.

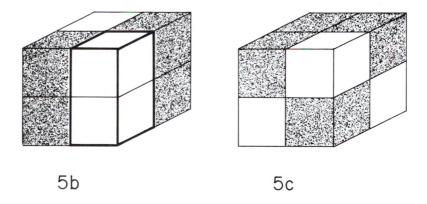

5b   5c

is fragile by

$$p_1 = p_0^8 + 8p_0^7(1 - p_0)$$
$$+ 16p_0^6(1 - p_0)^2 + 8p_0^5(1 - p_0)^3 + 2p_0^4(1 - p_0)^4$$
$$= p_0^4(3p_0^4 - 8p_0^3 + 4p_0^2 + 2) \tag{15.10}$$

After renormalization exactly the same statistics are applicable at higher orders. Thus we can write

$$p_{n+1} = p_n^4(3p_n^4 - 8p_n^3 + 4p_n^2 + 2) \tag{15.11}$$

If the characteristic size of the first-order cell is $2h$ then the characteristic size of the nth-order cell is $2^n h$.

Figure 15.6 shows the dependence of p_{n+1} on p_n given in (15.11). In order to consider the fixed points it is appropriate to rewrite (15.11) as

$$f(x) = x^4(3x^4 - 8x^3 + 4x^2 + 2) \tag{15.12}$$

The fixed points are obtained by setting $f(x) = x$ with the result

$$x = x^4(3x^4 - 8x^3 + 4x^2 + 2) \tag{15.13}$$

In the range $0 < x < 1$ there are three fixed points $x = 0, 0.896, 1$. The corresponding values of $\lambda = df/dx$ are 0, 1.766, 0. The fixed points at $x = 0$ and 1 are stable since $|\lambda| < 1$ but the fixed point at $x = 0.896$ is unstable.

In order to illustrate further the iteration given by (15.11) we consider a specific case. For $p_0 = 0.6$ we find $p_1 = 0.2736$, $p_2 = 0.0118$, and $p_3 = 3.91 \times 10^{-8}$ as illustrated in Figure 15.6. The construction is again the same as that used in Figures 10.1–10.5 and in Figure 15.3. As the iteration is continued to large n the probability p_n approaches the stable fixed point $p_\infty = 0$. Fragmentation does not occur for $p_0 = 0.6$. The unstable fixed point at $p^* = 0.896$ is a critical point corresponding to catastrophic fragmentation. For probabilities smaller than the critical value $0 < p_0 < p^*$ the iteration is to $p_\infty = 0$ and fragmentation does not occur. At the critical limit p^* the probability of fragmentation is the same at all orders. Thus it is

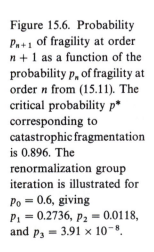

Figure 15.6. Probability p_{n+1} of fragility at order $n + 1$ as a function of the probability p_n of fragility at order n from (15.11). The critical probability p^* corresponding to catastrophic fragmentation is 0.896. The renormalization group iteration is illustrated for $p_0 = 0.6$, giving $p_1 = 0.2736$, $p_2 = 0.0118$, and $p_3 = 3.91 \times 10^{-8}$.

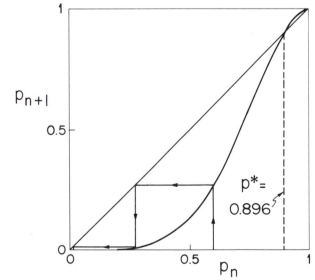

appropriate to set p^* equal to the probability f that a cell will fragment into eight elements, which was introduced in (3.22). Setting $f = p^* = 0.896$ and substituting into (3.28) we find that the resulting fractal dimension is $D = 2.84$.

It is of interest to relate the renormalization group approach to fragmentation as given above to the discussion of fragmentation given in Chapter 3. The renormalization group approach provides a rational basis for assuming that the fragmentation probability f is scale invariant. The applicable value of f is model dependent. For the 'pillar of strength' model the derived fractal dimension for fragments is $D = 2.84$. For the comminution model proposed in Chapter 3 the derived fractal dimension for fragments was $D = 2.6$. The latter is in somewhat better agreement with observed distributions.

As our final application of the renormalization group method we consider a fractal-tree model for the rupture of a fault. When a fault ruptures the result is an earthquake. An earthquake is in many ways analogous to a critical phenomenon. At a critical value of the stress a catastrophic event, the earthquake, occurs. As a result of the earthquake there is a discontinuity in the stress and strain.

A fault is generally considered to be a planar structure and can be modeled using a velocity-weakening coefficient of friction. An actual fault in the earth is generally more complex, being imbedded in a matrix that is fragmented by faults and joints. It appears appropriate to assume that a fault contains asperities on a wide range of scales. At a small scale an asperity can be an interacting roughness on the fault. At a large scale an asperity can be a bend in the fault. An asperity has a limiting strength and when this limiting strength is reached it will rupture. As a result of the rupture the stress on the asperity will be transferred to adjacent asperities and they may or may not rupture.

In many ways this rupture is analogous to the failure of a stranded cable. If one strand fails, the stress carried by that strand is transferred to adjacent unbroken strands. Extensive numerical calculations of this failure problem have been carried out (Harlow and Phoenix, 1982). Two observations are of interest. The first is that a stranded cable will fail after only a few strands have failed. The second is that the load that can be carried by a stranded cable is, on average, less than the load that could have been carried on a single-strand cable of the same mass. Both of these observations appear to be applicable to faults. There is seldom any seismic indication of small ruptures on a fault that is about to fail producing a large earthquake. Also, the failure stress on faults is generally less than that predicted by simple frictional considerations.

In order to model fault rupture Smalley *et al.* (1985) considered the failure of a fractal tree; they applied the renormalization group approach. Their basic model is illustrated in Figure 15.7(a). The force F is carried by each of the two first-order elements that make up the first-order cell. If one element fails the force on it is transferred to the other element and it must carry the force $2F$. A third-order fractal tree is illustrated in Figure 15.7(b). It includes four first-order cells each of which carries a force $2F$. These each become second-order elements in the two second-order cells each of which carries a force $4F$. The two second-order cells become third-order elements in the single third-order cell, which carries a force $8F$.

The objective is to determine the probability of failure of a cell in terms of the probabilities of failure of its elements. We assume that the probability $p_0(F)$ of failure of an element is given by the quadratic Weibull distribution (3.3):

$$p_0(F) = 1 - \exp\left[-\left(\frac{F}{F_0}\right)^2\right] \tag{15.14}$$

where F_0 is a reference strength.

For a cell containing two elements that are either broken (failed) or unbroken, four states are possible: (1) $[bb]$, (2) $[bu]$, (3) $[ub]$, (4) $[uu]$, where b represents a broken element and u represents an unbroken element. Note that states (2) and (3) are equivalent and can be combined into a single state with a multiplicity of two. The probabilities of these states in terms of the probability of failure p_0 are:

$$[bb] \qquad\qquad\qquad p_0^2 \tag{15.15}$$

$$[ub] \qquad\qquad\qquad 2p_0(1 - p_0) \tag{15.16}$$

Figure 15.7. Illustration of the fractal tree model for the renormalization group approach to an earthquake rupture. (a) The basic model; (b) a third-order construction.

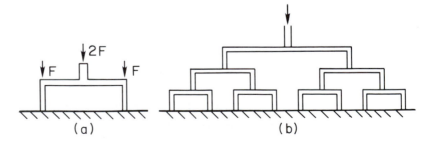

(a)   (b)

$$[uu] \qquad\qquad (1 - p_0)^2 \qquad\qquad (15.17)$$

In writing (15.15) to (15.17) the transfer of the force (stress) when an element fails has not been considered.

If one element fails and the other is unbroken it is necessary to determine whether the second element will fail when the stress from the first element is transferred to it. We introduce the conditional probability p_{21} that an unbroken element already supporting a force F will fail when an additional force F is transferred to it. This mechanism for stress transfer leads to induced failures. The probabilities that the $[ub]$ state will be broken or unbroken under stress transfer are given by:

$$[ub] \rightarrow [bb] \qquad\qquad 2p_0(1 - p_0)p_{21} \qquad\qquad (15.18)$$
$$[ub] \rightarrow [ub] \qquad\qquad 2p_0(1 - p_0)(1 - p_{21}) \qquad\qquad (15.19)$$

From (15.14) and (15.17) the probability that a zero-order cell fails, p_1, is given by

$$p_1 = p_0^2 + 2p_0(1 - p_0)p_{21} \qquad\qquad (15.20)$$

It is now necessary to determine the conditional probability p_{21}; it is given by

$$p_{21} = \frac{p_0(2F) - p_0(F)}{1 - p_0(F)} \qquad\qquad (15.21)$$

where $p_0(2F)$ is the probability of failure uder a force $2F$. For the quadratic Weibull distribution given in (15.14) we have

$$p_0(2F) = 1 - \exp\left[-\left(\frac{2F}{F_0}\right)^2 \right] \qquad\qquad (15.22)$$

The substitution of (15.14) into (15.22) gives

$$p_0(2F) = 1 - [1 - p_0(F)]^4 \qquad\qquad (15.23)$$

Combining (15.21) and (5.23) the conditional probability for the quadratic Weibull distribution is given by

$$p_{21} = 1 - (1 - p_0)^3 \qquad\qquad (15.24)$$

Substitution of (15.24) into (15.20) gives the probability that a cell fails, p_1, in terms of the probability that the element fails, p_0:

$$p_1 = 2p_0[1 - (1 - p_0)^4] - p_0^2 \qquad\qquad (15.25)$$

After renormalization exactly the same statistics are applicable to the second-order cell with the result

$$p_2 = 2p_1[1 - (1 - p_1)^4] - p_1^2 \qquad\qquad (15.26)$$

This result can be applied to the nth-order cell with the result

$$p_{n+1} = 2p_n[1 - (1 - p_n)^4] - p_n^2 \qquad (15.27)$$

Again this recursive relation resembles the logistic map considered in Chapter 10.

Figure 15.8 shows the dependence of p_{n+1} on p_n given in (15.27). In order to consider the fixed points it is appropriate to rewrite (15.27) as

$$f(x) = 2x[1 - (1 - x)^4] - x^2 \qquad (15.28)$$

The fixed points are obtained by setting $f(x) = x$ with the result

$$x = 2x[1 - (1 - x)^4] - x^2 \qquad (15.29)$$

In the range $0 < x < 1$ there are three fixed points $x = 0, 0.2063, 1$. The corresponding values of $\lambda = df/dx$ are 0, 1.619, 0. The fixed points at $x = 0$ and 1 are stable since $|\lambda| = 1$ but the fixed point at $x = 0.2063$ is unstable.

In order to illustrate further the iteration of the probabilities given by (15.27) we consider two specific cases. For $p_0 = 0.1$ we find $p_1 = 0.05878$, $p_2 = 0.02184$, and $p_3 = 0.00322$ as illustrated in Figure 15.8. The construction is the same as used in Figures 10.1–10.5. As the iteration is continued to large n the probability p_n approaches

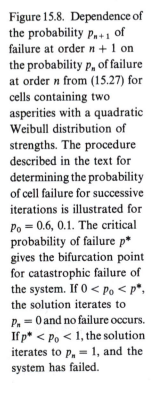

Figure 15.8. Dependence of the probability p_{n+1} of failure at order $n + 1$ on the probability p_n of failure at order n from (15.27) for cells containing two asperities with a quadratic Weibull distribution of strengths. The procedure described in the text for determining the probability of cell failure for successive iterations is illustrated for $p_0 = 0.6, 0.1$. The critical probability of failure p^* gives the bifurcation point for catastrophic failure of the system. If $0 < p_0 < p^*$, the solution iterates to $p_n = 0$ and no failure occurs. If $p^* < p_0 < 1$, the solution iterates to $p_n = 1$, and the system has failed.

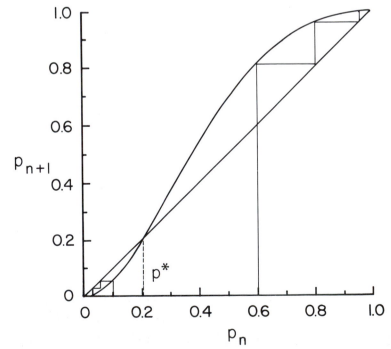

the stable fixed point $p_\infty = 0$. The large fractal tree does not fail for $p_0 = 0.1$. For $p_0 = 0.6$ we find $p_1 = 0.8093$, $p_2 = 0.9615$, and $p_3 = 0.9985$ as illustrated in Figure 15.6. As the iteration is continued to large n the probability p_n approaches the stable fixed point $p_\infty = 1$. The large fractal tree fails for $p_0 = 0.6$.

The unstable fixed point at $p^* = 0.2063$ is a critical point. For probabilities smaller than the critical value $0 < p_0 < p^*$ the iteration is to the unbroken limit $p_\infty = 0$. For probabilities greater than the critical value $p^* < p_0 < 1$ the iteration is to the broken limit $p_\infty = 1$. The value $p^* = 0.2063$ corresponds to the catastrophic failure of the fractal tree.

Because of the transfer of stress all elements fail when the probability of failure of an individual element is only 0.2063. From (15.14) this corresponds to $F/F_0 = 0.4807$. Considered individually half the elements will have failed when $p_0 = 0.5$ and $F/F_0 = 0.8326$. Thus the transfer of stress results in a lower failure stress than if a single element was considered.

Clearly this renormalization group approach is an idealization of actual faults. However, some of the results are directly related to field observations. A failure of a few elements cascades into a catastrophic total failure. Actual faults do not have enhanced seismicity prior to a major rupture. This observation coupled with the model studies indicate that stress transfer is important in fault ruptures. The relatively low failure stress is also in agreement with field observations.

Problems

Problem 15.1. A unit square is divided into 16 smaller squares of equal size and the four central squares are removed; the construction is repeated. Assume that the remaining squares represent a percolation cluster and determine n_e for $n = 16$ and 256.

Problem 15.2. Determine the equivalent expression to (15.2) for a cubic array; use the Menger sponge as an example.

Problem 15.3. Consider the 3×3 renormalization group approach to the two dimensional array of square elements. For this array the tabulation (permeable elements, impermeable elements, alternative configurations, and permeable configurations) is: (0, 9, 1, 0), (1, 8, 9, 0), (2, 7, 36, 0), (3, 6, 84, 3), (4, 5, 125, 21), (5, 4, 125, 58), (6, 3, 84, 67), (7, 2, 36, 36), (8, 1, 9, 9), (9, 0, 1, 1). Derive equation (15.8).

Problem 15.4. Consider the $2 \times 2 \times 2$ renormalization group approach to the three-dimensional array of cubic elements. For this array the tabulation (permeable elements, impermeable elements, alternative configurations, and permeable configurations) is: (0, 8, 1, 0), (1, 7, 8, 0), (2, 6, 28, 4), (3, 5, 56, 24), (4, 4, 70, 54), (5, 3, 56, 56), (6, 2, 28, 28), (7, 1, 8, 8), (8, 0, 1, 1). Derive equation (15.9).

Problem 15.5. Assuming the dark elements in Figure 15.1 are permeable what is the size of the largest percolation cluster?

Problem 15.6. Assuming the light elements in Figure 15.1 are permeable what is the size of the largest percolation cluster?

Problem 15.7. Derive the conditional probability given by (15.21).

Problem 15.8. Derive (15.23) from (15.14) and (15.22).

Problem 15.9. Consider the third-power Weibull distribution given by

$$p_0(F) = 1 - \exp\left[-\left(\frac{F}{F_0}\right)^3\right]$$

instead of (15.14). Show that (15.23) should be replaced by

$$p_0(2F) = 1 - [1 - p_0(F_0)]^8$$

Show that the recursive failure relation becomes

$$p_{n+1} = 2p_n - [1 - (1 - p_n)^8] - p_n^2$$

Find the value of the unstable fixed point p^* and the corresponding value of F/F_0.

Problem 15.10. Consider the fourth-power Weibull distribution given by

$$p_0(F) = 1 - \exp\left[-\left(\frac{F}{F_0}\right)^4\right]$$

instead of (15.14). Show that (15.23) should be replaced by

$$p_0(2F) = 1 - [1 - p_0(F)]^{16}$$

Show that the recursive failure relation becomes

$$p_{n+1} = 2p_n[(1 - p_n)^{16}] - p_n^2$$

Find the value of the unstable fixed point p^* and the corresponding value of F/F_0.

CHAPTER SIXTEEN

Self-organized criticality

In the last chapter we considered the renormalization group method for treating large interactive systems. By assuming scale invariance a relatively small system could be scaled upwards to a large interactive system. The approach is often applicable to systems that have critical point phenomena. In this chapter we consider an alternative approach to large interactive systems. This approach is called self-organized criticality. A system is said to be in a state of self-organized criticality if it is maintained near a critical point (Bak *et al.*, 1988). According to this concept a natural system is in a marginally stable state; when perturbed from this state it will evolve naturally back to the state of marginal stability. In the critical state there is no longer a natural length scale so that fractal statistics are applicable.

The simplest physical model for self-organized criticality is a sand pile. Consider a pile of sand on a circular table. Grains of sand are randomly dropped on the pile until the slope of the pile reaches the critical angle of repose. This is the maximum slope that a granular material can maintain without additional grains sliding down the slope. One hypothesis for the behavior of the sand pile would be that individual grains could be added until the slope is everywhere at an angle of repose. Additional grains would then simply slide down the slope. This is not what happens. The sand pile never reaches the hypothetical critical state. As the critical state is approached additional sand grains trigger landslides of various sizes. The frequency–size distribution of landslides is fractal. The sand pile is said to be in a state of self-organized criticality. On average the number of sand grains added balances the number that slide down the slope and off the table. But the actual number of grains on the table fluctuates continuously.

The principles of self-organized criticality are illustrated using a simple cellular-automata model. As in the previous chapter we again consider a square grid of *n* boxes. Particles are added to and lost from the grid using the following procedure.

(1) A particle is randomly added to one of the boxes. Each box on the grid is assigned a number and a random-number generator is used to determine the box to which a particle is added. This is a statistical model.

(2) When a box has four particles it is unstable and the four particles are redistributed to the four adjacent boxes. If there is no adjacent box the particle is lost from the grid. Redistributions from edge boxes result in the loss of one particle from the grid. Redistributions from the corner boxes result in the loss of two particles from the grid.

(3) If after a redistribution of particles from a box any of the adjacent boxes has four or more particles, it is unstable and one or more further redistributions must be carried out. Multiple events are common occurrences for large grids.

(4) The system is in a state of marginal stability. On average added particles must be lost from the sides of the grid.

This is a nearest neighbor model. At any one step a box interacts only with its four immediate neighbors. However, in a multiple event interactions can spread over a large fraction of the grid.

The behavior of the system is characterized by the statistical frequency–size distribution of events. The size of a multiple event can be quantified in several ways. One measure is the number of boxes that become unstable in a multiple event. Another measure is the number of particles lost from the grid during a multiple event.

When particles are first added to the grid there are no redistributions and no particles are lost from the grid. Eventually the system reaches a quasi-equilibrium state. On average the number of particles lost from the edges of the grid is equal to the number of particles added. Initially, small redistribution events dominate but in the quasi-equilibrium state the frequency–size distribution is fractal. This is the state of self-organized criticality. There is a strong resemblance to the renormalization group approach considered in the last chapter. In the renormalization group approach the frequency–size statistics are only fractal at the critical point. In the cellular automata model the frequency–size statistics are fractal only in the state of self-organized criticality.

The behavior of a sand pile and the behavior of the cellular automata model have remarkable similarities to the seismicity associated with an active tectonic zone. The addition of particles to the grid is analogous to the addition of stress caused by the relative displacement between two surface plates, say, across the San Andreas fault. The multiple events in which particles are transferred and are lost from the grid are analogous to earthquakes in which some accumulated stress is transferred and some is lost. There is a strong similarity between the frequency–magnitude statistics of multiple events and the Gutenberg–Richter statistics for earthquakes. Before considering the analogy further we will describe the behavior of the cellular automata model in some detail.

As a specific example we consider the 3×3 grid illustrated in Figure 16.1. The nine boxes are numbered sequentially from left to right and top to bottom as illustrated in Figure 16.1(a). The cellular automata model has been run for some time to establish a state of self-organized criticality. The further evolution of the model is as follows and is illustrated in Figure 16.1(b).

Step 1 A particle has been randomly added to box 8. The number of particles in this box has been increased from two to three.

Step 2 A particle has been randomly added to box 6 increasing the

Figure 16.1. Illustration of the cellular automata model for a 3×3 grid of boxes. The boxes are numbered 1 to 9 as shown in (a). Particles are randomly added to boxes in (b) as shown in steps 1 and 2. In step 3a an added particle in box 5 gives four particles and these are redistributed to the adjacent boxes. Nine more redistributions are required in steps 3b to 3j before the grid is stabilized. The first number below the grid is the number of boxes that have been unstable in the sequence of redistributions. The second number is the cumulative number of particles that have been lost from the grid in the sequence of redistributions.

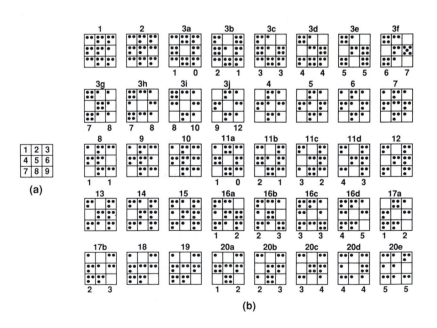

(a)

(b)

number of particles from one to two. This addition is illustrated in the change between steps 1 and 2 in Figure 16.1(b).

Step 3a A particle has been randomly added to box 5 increasing the number of particles from three to four and making it unstable; the four particles are redistributed to the four adjacent boxes increasing the number of particles in box 2 from three to four, the number of particles in box 4 from three to four, the number of particles in box 6 from two to three, and the number of particles in box 8 from three to four. Boxes 2, 4, and 8 are now unstable. No particles are lost from the grid. This redistribution is illustrated in step 3a in Figure 16.1(b). The numbers below the grid are, on the left, the cumulative numbers of boxes subject to redistribution and, on the right, the cumulative number of particles lost from the grid.

Step 3b Since several boxes are now unstable an arbitrary choice must be made about which box will be considered first for further redistribution. The choice does not have a significant effect on the statistical evolution of the system. The four particles in box 2 are redistributed. One is lost from the grid and box 3 becomes unstable with four particles. Boxes 3, 4, and 8 remain unstable. In this sequence of redistributions two boxes have been unstable and one particle has been lost from the grid.

Step 3c The four particles in box 3 are redistributed. Two are lost from the grid and box 6 becomes unstable with four particles. Boxes 4, 6, and 8 remain unstable. In this sequence of redistributions three boxes have been unstable and three particles have been lost from the grid.

Step 3d The four particles in box 4 are redistributed. One is lost from the grid and box 1 becomes unstable with four particles. Boxes 1, 6, and 8 remain unstable. In this sequence of redistributions four boxes have been unstable and four particles have been lost from the grid.

Step 3e The four particles on grid point 8 are redistributed. One is lost from the grid and boxes 7 and 9 become unstable with four particles. Boxes 1, 6, 7, and 9 remain unstable. In this sequence of redistributions five boxes have been unstable and five particles have been lost from the grid.

Step 3f The four particles in box 9 are redistributed. Two are lost from the grid and box 6 is now unstable with five particles. Grid points 1, 6, and 7 remain unstable. In this sequence of redistributions six boxes have been unstable and seven particles have been lost from the grid.

Step 3g Four of the five particles in box 6 are redistributed. One is lost from the grid and box 5 is now unstable. Boxes 1, 5, and 7 remain unstable. In this sequence of redistributions seven boxes have been unstable and eight particles have been lost from the grid.

Step 3h The four particles in box 5 are redistributed for the second time. No particles are lost and no boxes are made unstable. Boxes 1 and 7 remain unstable. In this sequence of redistributions seven boxes have been unstable and eight particles have been lost from the grid.

Step 3i The four particles in box 7 are redistributed and two are lost from the grid. No boxes are made unstable so that 1 is the only remaining unstable box. In this sequence of redistributions eight boxes have been made unstable and ten particles have been lost from the grid.

Step 3j The four particles in box 1 are redistributed and two are lost from the grid. No boxes remain unstable so that the sequence of 10 redistributions has completed step 3. During step 3 all nine boxes were unstable and 12 particles were lost from the grid.

Step 4 A particle has been randomly added to box 5 increasing the number of particles from zero to one.

Step 5 A particle has been randomly added to box 6 increasing the number of particles from two to three.

This relatively simple example illustrates how the cellular automata model works. In order to develop significant statistics larger grids must be considered. Kadanoff *et al.* (1989) have carried out extensive studies of the behavior of this model as a function of grid size. One statistical measure of the size of an event is the number of grid points that become unstable. The results for a 50×50 grid of boxes are given in Figure 16.2. The number of events N in which a specified number of boxes n participated is given as a function of the number of boxes. A good correlation with a fractal power law is obtained, with a slope of 1.03. Since the number of grid points is equivalent to an area the equivalent fractal dimension is $D = 2.06$. This statistical behavior appears to resemble that of distributed seismicity. However, the statistics in Figure 16.2 are not cumulative. In fact a fractal relation is not obtained for the cumulative statistics.

Slider-block models that exhibit self-organized criticality can also be constructed. In Chapter 11 we showed that a pair of interacting slider blocks can exhibit deterministic chaos. This model is easily extended to include large numbers of slider blocks (Carlson and Langer, 1989; Bak and Tang, 1989; Takayasu and Matsuzaki, 1988; Nakanishi, 1990; Ito and Matsuzaki, 1990; Sornette and Sornette, 1989, 1990). The slider blocks interact only with their nearest neighbors. When a block slips a fraction of the stored energy is lost due to friction and a fraction is transferred to the nearest neighbors. This transfer can lead to a multiple-block failure. The frequency–size distribution of multiple block failures resemble those for the cellular automata model given above and the Gutenberg–Richter relationship for earthquakes. It should be noted that there is one important difference between the cellular automata model and the multiple slider-block model. The cellular automata model is statistical while the slider-block model is completely deterministic.

The frequency–size distribution of events associated with self-organized criticality certainly resembles the regional distribution of earthquakes in a zone of active tectonics. This suggests that interactions between faults play an essential role in the behavior of such zones.

An important consequence of a critical state in the crust is the large range of interactions. A basic question is whether an earthquake on one part of the planet, say Mexico, can trigger an earthquake at a large distance, say Japan. The classical approach to earthquakes would say that this is impossible. The stresses associated with seismic waves are too small to trigger an earthquake and there is no observational evidence for triggered events on this spatial scale. The stress changes associated with the fault displacement are localized

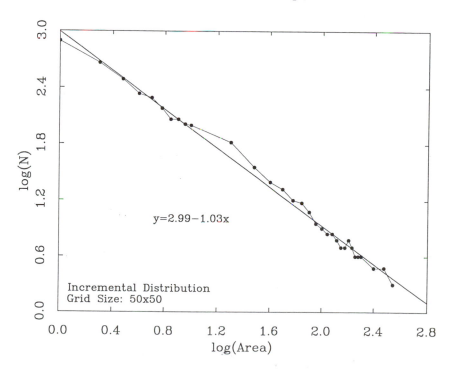

Figure 16.2. Statistics for a cellular automata model on a 50 × 50 grid. The number N of events in which a specified number A of boxes became unstable is given as a function of A.

and are damped by the athenospheric viscosity. However, interactions at large distances are a characteristic of critical phenomena. The interactions are not through the direct transmission of stress but through the interactions of faults with each other.

This action at a distance may help to explain the apparent success of the earthquake prediction algorithms developed at the International Institute for the Theory of Earthquake Prediction and Theoretical Geophysics in Moscow under the direction of Academician V. I. Kellis-Borok. This approach is based on pattern recognition of distributed regional seismicity (Kellis-Borok, 1990; Kellis-Borok and Rotwain, 1990; Kellis-Borok and Kossobokov, 1990). The pattern recognition includes quiescence (Schreider, 1990), increases in the clustering of events, and changes in aftershock statistics (Molchan *et al.*, 1990). In reviewing regional seismicity after a major earthquake it is often observed that the regional seismicity in the vicinity of the fault rupture was anomalously low for several years prior to the earthquake (Kanamori, 1981; Wyss and Haberman, 1988). This is known as seismic quiescence. The problem has been to provide quantitative measures of quiescence prior to the major earthquake. We discussed the fractal clustering of earthquakes in Chapter 6. The clustering of regional seismicity appears to become more fractal-like prior to a large earthquake. There also appears to be a systematic reduction in the number of aftershocks associated with regional intermediate-sized earthquakes prior to a major earthquake. Algorithms were developed to search earthquake catalogues for anomalous statistics over areas with diameters of 500 km. When a threshold of the anomalous behavior was reached a warning of the time of increased probability (TIP) of an earthquake was issued.

On a world-wide basis TIPs were triggered prior to 42 of 47 events. TIPs were released prior to the Armenian earthquake on 7 December 1988 and to the Loma Prieta earthquake on 17 October 1989. These are illustrated in Figure 16.3. The TIP issued for region 3 in the Caucasus during January 1987 was still in effect when the Armenian earthquake occurred in this region on 7 December 1988. TIPs were issued for region 5 in California during October 1984 and for region 6 during January 1985. These warnings were still in effect when the Loma Prieta earthquake occurred within both these overlapping regions on 17 October 1989.

The fault rupture of the Loma Prieta earthquake extended over about 40 km. However, the prediction algorithms detected anomalous seismic behavior over two regions with diameters of 500 km. Self-organized criticality can explain anomalous correlated behavior over relatively large distances.

(a)

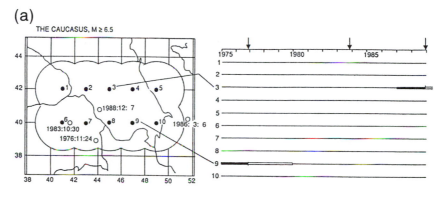

Figure 16.3. Illustrations of the Armenian (7 December 1988) and Loma Prieta (18 October 1989, Moscow time) earthquakes by Keilis-Borok (1990). In (a) the Caucasus region is broken up into 10 areas with diameters of 500 km; two warnings (for region 3 and 9) are shown on the right. The locations and times of four earthquakes are also given. In (b) the California–Nevada region is broken up into eight areas with diameters of 500 km. Four warnings (for regions 4–6, 8) and the locations and times of four earthquakes are given.

(b)

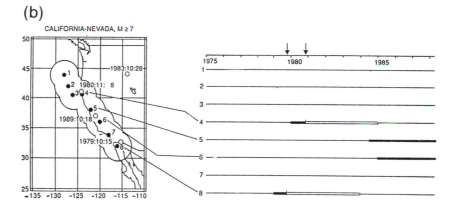

Problems

Problem 16.1. Consider the evolution of the cellular automata model illustrated in Figure 16.1(b). (a) Which boxes have an additional particle in steps 6, 7, 9 and 10? (b) Which boxes are unstable and how many particles are lost from the grid in steps 8, 11a, 11b, 11c, and 11d?

Problem 16.2. Consider the evolution of the cellular automata model illustrated in Figure 16.1(b). (a) Which boxes have an additional particle in steps 12, 13, 14, and 15? (b) Which boxes are unstable and how many particles are lost from the grid in steps 16a, 16b, 16c, 16d, 17a, and 17b?

Problem 16.3. Consider the evolution of the cellular automata model illustrated in Figure 16.1(b). (a) Which boxes have an additional particle in steps 18 and 19? (b) Which boxes are unstable and how many particles are lost from the grid in steps 20a, 20b, 20c, 20d, and 20e?

Problem 16.4.

$$\begin{array}{cc} 1 & 2 \\ 3 & 4 \end{array}$$

4332413323143421132121324323

(a)   *(b)*

Consider a 2×2 grid of four boxes as illustrated above in (*a*). Also given above in (*b*) is a sequence of random numbers in the range 1–4. Use the random numbers to assign particles to boxes and carry out the cellular automata model described in this chapter.

Problem 16.5.

1234

Consider the linear grid of four boxes illustrated above. Use the sequence of random numbers given in Problem 16.4 to assign particles to the four boxes. Use the following rules: When a box has two particles it is unstable and they are redistributed to the two adjacent boxes. If either of these boxes has two elements they are again redistributed. Particles are lost from the ends of the linear grid.

CHAPTER SEVENTEEN

Where do we stand?

The concepts of fractals and chaos were introduced in 1967 (Mandelbrot, 1967) and in 1963 (Lorenz, 1963), respectively. Unlike many advances in science, the attribution is in both cases quite clear. In both cases it took more than 10 years before either concept received wide attention. In fact, today, many scientists regard fractals and chaos as passing fads.

We are at a critical phase in the development of both concepts. Only time will tell whether either or both represent major scientific breakthroughs. It will be the generations of scientists who are exposed to this book and other current publications who will determine whether the concepts will flourish or die.

There is no question that fractals are a useful empirical tool. They provide a rational means for the extrapolation and interpolation of observations. But do they have a substantive basis for their applications? Chaos and related concepts such as self-organized criticality may provide this basis but much work remains to be done.

More and more systems are found that exhibit deterministic chaos. But these systems are generally low-order analogs to systems of practical importance. Experimental systems can be constructed that behave in a chaotic manner but these are far removed from practical problems. Major objectives are to bridge the gap between the Lorenz equations and turbulence, to bridge the gap between slider-block models and earthquakes, and to bridge the gap between the Rikitake dynamo and the generation of the earth's magnetic field. These are just a few of the challenges for the next generation, and quite likely for succeeding generations, of scientists.

References

Aki, K. (1981). A probabilistic synthesis of precursory phenomena, in *Earthquake Prediction*, D. W. Simpson & P. G. Richards, eds, pp. 566–74, American Geophysical Union, Washington, D.C.

Allègre, C. J., Le Mouel, J. L. & Provost, A. (1982). Scaling rules in rock fracture and possible implications for earthquake prediction, *Nature* **297**, 47–9.

Anderson, J. G. (1986). Seismic strain rates in the central and eastern United States, *Seis. Soc. Am. Bull.* **76**, 273–90.

Anderson, J. G. & Luco, L. E. (1983). Consequences of slip rate constraints on earthquake occurrence relations, *Seis. Soc. Am. Bull.* **73**, 471–96.

Andrews, D. J. & Hanks, T. C. (1985). Scarp degraded by linear diffusion: inverse solution for age, *J. Geophys. Res.* **90**, 10193–208.

Bak, P. & Tang, C. (1989). Earthquakes as a self-organized critical phenomenon, *J. Geophys. Res.* **94**, 15635–7.

Bak, P., Tang, C. & Wiesenfeld, K. (1988). Self-organized criticality, *Phys. Rev.* **A38**, 364–74.

Barenblatt, G. I., Zhivago, A. V., Neprochnov, Y. P. & Ostrovskiy, A. A. (1984). The fractal dimension: A quantitative characteristic of ocean-bottom relief, *Oceanology* **24**, 695–7.

Barton, C. C. & Hsieh, P. A. (1989). *Physical and Hydrologic-Flow Properties of Fractures*, 28th International Geological Congress Field Trip Guidebook T385, American Geophysical Union, Washington, D.C., p. 36.

Bell, T. H. (1975). Statistical features of sea-floor topography, *Deep-sea Res.* **22**, 883–92.

Bell, T. H. (1979). Messoscale sea floor roughness, *Deep-sea Res.* **26A**, 65–76.

Bennett, J. G. (1936). Broken coal, *J. Inst. Fuel* **10**, 22–39.

Bergé, P., Poneau, Y. & Vidal, C. (1986). *Order within Chaos*, John Wiley and Sons, New York, 329 pp.

Berkson, J. M. & Mathews, J. E. (1983). Statistical properties of sea-floor roughness, in *Acoustics and the Sea-bed*, N. G. Pace, ed., pp. 215–23, Bath University Press, Bath.

Biegel, R. L., Sammis, C. G. & Dieterich, J. H. (1989). The frictional properties of a simulated gouge having a fractal particle distribution. *J. Struct. Geol.* **11**, 827–46.

Brown, S. R. (1987). A note on the description of surface roughness using fractal dimension, *Geophys. Res. Lett.* **14**, 1095–8.

Brown, S. R. & Scholtz, S. H. (1985). Broad bandwidth study of the topography of natural rock surfaces, *J. Geophys. Res.* **90**, 12575–82.

Bucknam, R. C. & Anderson, R. E. (1979). Estimation of fault-scarp ages from a scarp-height–slope-angle relationship, *Geology* **7**, 11–14.

Burridge, R. & Knopoff, L. (1967). Model and theoretical seismicity, *Seis. Soc. Am. Bull.* **57**, 341–71.

Byerlee, J. (1978). Friction of rocks, *Pure Appl. Geophys.* **116**, 615–26.

Cao, T. & Aki, K. (1984). Seismicity simulation with a mass–spring model and a displacement hardening–softening friction law, *Pure Appl. Geophys.* **122**, 10–24.

Cao, T. & Aki, K. (1986). Seismicity simulation with a rate-and-state friction law, *Pure Appl. Geophys.* **124**, 487–513.

Cargill, S. M., Root, D. H. & Bailey, E. H. (1980). Resources estimation from historical data: Mercury, a test case, *J. Int. Assoc. Math. Geol.* **12**, 489–522.

Cargill, S. M., Root, D. H. & Bailey, E. H. (1981). Estimating usable resources from historical industry data, *Econ. Geol.* **76**, 1081–95.

Carlson, J. M. & Langer, J. S. (1989). Mechanical model of an earthquake fault, *Phys. Rev.* **A40**, 6470–84.

Chepil, W. S. (1950). Methods of estimating apparent density of discrete soil grains and aggregates, *Soil Sci.* **70**, 351–62.

Chinnery, M. A. (1979). A comparison of the seismicity of three regions of the eastern US, *Seis. Soc. Am. Bull.* **69**, 757–72.

Clark, G. B. (1987). *Principles of Rock Fragmentation*, John Wiley and Sons, New York, 610 pp.

Cohen, S. (1977). Computer simulation of earthquakes, *J. Geophys. Res.* **82**, 3781–96.

Cook, A. E. & Roberts, P. H. (1970). The Rikitake two-disc dynamo system, *Proc. Camb. Phil. Soc.* **68**, 547–69.

Culling, W. E. H. (1960). Analytical theory of erosion, *J. Geol.* **68**, 336–44.

Culling, W. E. H. (1963). Soil creep and the development of hillside slopes, *J. Geol.* **71**, 127–61.

Culling, W. E. H. (1965). Theory of erosion on soil-covered slopes, *J. Geol.* **73**, 230–54.

Curran, D. R., Shockey, D. A., Seaman, L. & Austin, M. (1977). Mechanisms and models of cratering in earth media, in *Impact and Explosion Cratering*, D. J. Roddy, R. O. Pepin & R. P. Merrill, eds, pp. 1057–87, Pergamon Press, New York.

Daccord, G. & Lenormand, R. (1987). Fractal patterns from chemical dissolution, *Nature* **325**, 41–3.

Dewey, J. F. (1975). Finite plate implications: Some implications for the evolution of rock masses at plate margins, *Am. J. Sci.* **275A**, 260–84.

De Wijs, H. J. (1951). Statistics of ore distribution. Part I: Frequency distribution of assay values, *Geol. Mijnb.* **13**, 365–75.

De Wijs, H. J. (1953). Statistics of ore distribution. Part II: Theory of binomial distribution applied to sampling and engineering problems, *Geol. Mijnb.* **15**, 12–24.

Dieterich, J. H. (1972). Time-dependent friction as a possible mechanism for aftershocks, *J. Geophys. Res.* **77**, 3771–81.

Dieterich, J. H. (1981). Constructive properties of faults with simulated gouge, in *Mechanical Behavior of Crustal Rocks*, N. L. Carter, N. Friedman, J. M. Logan & D. W. Steans, eds, pp. 103–20, American Geophysical Union, Geophysical Monographs 24.

Drew, L. J., Schuenemeyer, J. H. & Bawiec, W. J. (1982). Estimation of the future rates of oil and gas discoveries in the western gulf of Mexico, US Geological Survey Professional Paper 1252, 26 pp.

Dubuc, B., Quiniou, J. F., Roques-Carmes, C., Tricot, C. & Zucker, S. W. (1989a). Evaluating the fractal dimension of profiles, *Phys. Rev.* **A39**, 1500–12.

Dubuc, B., Zucker, S. W., Tricot, C., Quiniou, J. F. & Wehbi, D. (1989b). Evaluating the fractal dimension of surfaces, *Proc. Roy. Soc. London* **A425**, 113–27.

Dziewonski, A. M., Ekstrom, G., Woodhouse, J. H. & Zwart, G. (1989). Centroid-moment tensor solutions for October–December 1987, *Phys. Earth Planet. Int.* **54**, 10–21.

Evernden, J. F. (1970). Study of regional seismicity and associated problems, *Seis. Soc. Am. Bull.* **60**, 393–446.

Feder, J. (1988). *Fractals*, Plenum Press, New York, p. 283.

Fluigeman, R. H. & Snow, R. S. (1989). Fractal analysis of long-range paleoclimatic data: Oxygen isotope record of Pacific Core V28-313, *Pure Appl. Geophys.* **131**, 307–13.

Fox, C. G. (1989). Empirically derived relationships between fractal dimension and power law from frequency spectra, *Pure Appl. Geophys.* **131**, 307–13.

Fox, C. G. & Hayes, D. E. (1985). Quantitative methods for analysing the roughness of the seafloor, *Rev. Geophys.* **23**, 1–48.

Fujiwara, A., Kamimoto, G. & Tsukamoto, A. (1977). Destruction of basaltic bodies by high-velocity impact, *Icarus* **31**, 277–88.

Gilbert, L. E. (1989). Are topographic data sets fractal?, *Pure Appl. Geophys.* **131**, 241–54.

Gilbert, L. E. & Malinverno, A. (1988). A characterization of the spectral density of residual ocean floor topography, *Geophys. Res. Let.* **15**, 1401–4.

Goodchild, M. F. (1980). Fractals and the accuracy of geographical measures, *Math. Geol.* **12**, 85–98.

Grady, D. E. & Kipp, M. E. (1987). Dynamic rock fragmentation, in *Fracture Mechanics of Rock*, B. K. Atkinson, ed., pp. 429–75, Academic Press, London.

Gu, J. C., Rice, J. R., Ruina, A. L. & Tse, S. T. (1984). Slip motion and stability of a single degree of freedom elastic system with rate and state dependent friction, *J. Phys. Solids* **32**, 167–96.

Gutenberg, B. & Richter, C. F. (1954). *Seismicity of the Earth and Associated Phenomenon*, 2nd edition, Princeton University Press, Princeton.

Hack, J. T. (1957). Studies of longitudinal stream profiles in Virginia and Maryland, U.S. Geological Survey Professional Paper 294B, 97 pp.

Hanks, T. C. & Kanamori, H. (1979). A moment–magnitude scale, *J. Geophys. Res.* **84**, 2348–50.

Hanks, T. C. & Wallace, R. E. (1985). Morphological analysis of the Lake Lahontan shoreline and beachfront fault scarps, Pershing County, Nevada, *Seis. Soc. Am. Bull.* **75**, 835–46.

Hanks, T. C., Buckman, R. C., Lajoie, K. R. and Wallace, R. E. (1984). Modification of wave-cut and faulting-controlled landforms, *J. Geophys. Res.* **89**, 5771–90.

Harlow, D. G. & Phoenix, S. L. (1982). Probability distributions for the strength of fibrous materials under load sharing, 1. Two-level failure and edge effects, *Adv. Appl. Prob.* **14**, 68–94.

Harris, D. P. (1984). *Mineral Resources Appraisal*, Oxford University Press, Oxford.

Hartmann, W. K. (1969). Terrestrial, lunar, and interplanetary rock fragmentation, *Icarus* **10**, 201–13.

Hewett, T. A. (1986). Fractal distributions of reservoir heterogeneity and their influence on fluid transport, Society of Petroleum Engineers Paper 15386, 15 pp.

Hirata, T. (1989a). Fractal dimension of fault systems in Japan: Fractal structure in rock fracture geometry at various scales, *Pure Appl. Geophys.* **131**, 157–70.

Hirata, T. (1989b). A correlation between the *b* value and the fractal dimension of earthquakes, *J. Geophys. Res.* **94**, 7507–14.

Hirata, T., Satoh, T. & Ito, K. (1987). Fractal structure of spatial distribution of microfracturing in rock, *Geophys. J. Roy. Astron. Soc.* **90**, 369–74.

Huang, J. & Turcotte, D. L. (1988). Fractal distributions of stress and strength and variations of *b*-value, *Earth Planet. Sci. Lett.* **91**, 223–30.

Huang, J. & Turcotte, D. L. (1989). Fractal mapping of digitized images: Application to the topography of Arizona and comparisons with synthetic images, *J. Geophys. Res.* **94**, 7491–5.

Huang, J. & Turcotte, D. L. (1990a). Are earthquakes an example of deterministic chaos?, *Geophys. Res. Lett.* **17**, 223–6.

Huang, J. & Turcotte, D. L. (1990b). Fractal image analysis: Application to the topography of Oregon and synthetic images, *J. Opt. Soc. Am.* **A7**, 1124–30.

Huang, J. & Turcotte, D. L. (1990c). Evidence for chaotic fault interaction in the seismicity of the San Andreas fault and Nakai trough, *Nature* **348**, 234–6.

Hyndman, R. D. & Weichert, D. H. (1983). Seismicity and rates of relative motion on the plate boundaries of western North America, *Geophys. J. Roy. Astron. Soc.* **72**, 59–82.

Ito, K. & Matsuzaki, M. (1990). Earthquakes as self-organized critical phenomena, *J. Geophys. Res.* **95**, 6853–68.

Ivanhoe, L. F. (1976). Oil/gas potential in basins estimates, *Oil Gas J.* **74**, Dec. 6, 154–5.

Johnston, A. C. & Nava, S. J. (1985). Recurrence rates and probability estimates for the New Madrid seismic zone, *J. Geophys. Res.* **90**, 6737–53.

Jurdy, D. M. & Stefanick, M. (1990). Models for the hotspot distribution, *Geophys. Res. Lett.* **17**, 1965–8.

Kadanoff, L. P., Nagel, S. R., Wu, L. & Zhou, S. M. (1989). Scaling and universality in avalanches, *Phys. Rev.* **A39**, 6524–33.

Kanamori, H. (1978). Quantification of earthquakes, *Nature* **271**, 411–14.

Kanamori, H. (1981). The nature of seismicity patterns before large earthquakes, in *Earthquake Prediction*, D. W. Simpson and P. G. Richards, eds, pp. 1–19, American Geophysical Union, Washington DC.

Kanamori, H. & Anderson, D. L. (1975). Theoretical basis of some empirical relations in seismology, *Seis. Soc. Am. Bull.* **65**, 1073–96.

Katz, A. J. & Thompson, A. H. (1985). Fractal sandstone pores: Implications for conductivity and pore formation, *Phys. Rev. Lett.* **54**, 1325–8.

Kellis-Borok, V. I. (1990). The lithosphere of the earth as a nonlinear system with implications for earthquake prediction. *Rev. Geophys.* **28**, 19–34.

Kellis-Borok, V. I. & Kossobokov, V. G. (1990). Premonitory activation of earthquake flow. Algorithm M8, *Phys. Earth Planet Int.* **61**, 73–83.

Kellis-Borok, V. I. & Rotwain, I. M. (1990). Diagnosis of time of increased probability of strong earthquakes in different regions of the world: Algorithm CN, *Phys. Earth Planet. Int.* **61**, 57–72.

Kenyon, P. M. & Turcotte, D. L. (1985). Morphology of a delta prograding by bulk sediment transport, *Geol. Soc. Am. Bull.* **96**, 1457–65.

King. G. (1983). The accommodation of large strains in the upper lithosphere of the earth and other solids by self-similar fault systems: The geometrical origin of *b*-value, *Pure. Appl. Geophys.* **121**, 761–815.

King, G. C. P. (1986). Speculations on the geometry of the initiation and termination processes of earthquake rupture and its relation to morphology and geological structure, *Pure Appl. Geophys.* **124**, 567–85.

King, G. C. P., Stein, R. S. & Rundle, J. B. (1988). The growth of geological structures by repeated earthquakes 1. Conceptual framework, *J. Geophys. Res.* **93**, 13307–18.

Kolmogorov, A. (1941). The local structure of turbulence in incompressible viscous fluid for very large Reynold's numbers, *Comptes Rendus, Doklady de l'Academie des Sciences de l'USSR* **30**, 301–5.

Krohn, C. E. (1988a). Fractal measurements of sandstones, shales, and carbonates, *J. Geophys. Res.* **93**, 3297–305.

Krohn, C. E. (1988b). Sandstone fractal and Euclidean pore volume distributions, *J. Geophys. Res.* **93**, 3286–96.

Krohn, C. E. & Thompson, A. H. (1986). Fractal sandstone pores: Automated measurements using scanning-electron microscope images, *Phys. Rev.* **B33**, 6366–74.

Landau, L. D. & Lifshitz, E. M. (1959). *Fluid Mechanics*, Pergamon Press, Oxford.

Lasky, S. G. (1950). How tonnage and grade relations help predict ore reserves, *Eng. Mining J.* **151**, 81–5.

Lorenz, E. N. (1963). Deterministic nonperiodic flow, *J. Atmos. Sci.* **20**, 130–41.

Madden, T. R. (1983). Microcrack connectivity in rocks: A renormalization group approach to the critical phenomena of conduction and failure in crystalline rocks, *J. Geophys. Res.* **88**, 585–92.

Main, I. G. & Burton, P. W. (1986). Long-term earthquake recurrence constrained by tectonic seismic moment release rates, *Seis. Soc. Am. Bull.* **76**, 297–304.

Malinverno, A. (1989). Testing linear models of sea-floor topography, *Pure Appl. Geophys.* **131**, 139–55.

Mandelbrot, B. B. (1967). How long is the coast of Britain? Statistical self-similarity and fractional dimension, *Science* **156**, 636–8.

Mandelbrot, B. B. (1975). Stochastic models for the earth's relief, the shape and the fractal dimension of the coastlines, and the number–area rule for islands, *Proc. Nat. Acad. Sci. USA* **72**, 3825–8.

Mandelbrot, B. B. (1982). *The Fractal Geometry of Nature*, Freeman, San Francisco.

Mandelbrot, B. B. (1985). Self-affine fractals and fractal dimension, *Physica Scripta* **32**, 257–60.

Mareschal, J. C. (1989). Fractal reconstruction of sea-floor topography, *Pure Appl. Geophys.* **131**, 197–210.

May, R. M. (1976). Simple mathematical models with very complicated dynamics, *Nature* **261**, 459–67.

Mayer, L. (1984). Dating Quaternary fault scarps formed in alluvium using morphologic parameters, *Quat. Res.* **22**, 300–13.

McClelland, L., Simkin, T., Summers, M., Nielson, E. & Stein, T. C., eds, (1989). *Global Volcanism 1975–1985*, Prentice Hall, Englewood Cliffs, NJ, 655 pp.

Molchan, G. M., Dmitrieva, O. E., Rotwain, I. M. & Dewey, J. (1990). Statistical analysis of the results of earthquake prediction, based on bursts of aftershocks, *Phys. Earth Planet. Int.* **61**, 128–39.

Molnar, P. (1979). Earthquake recurrence intervals and plate tectonics, *Seis. Soc. Am. Bull.* **69**, 115–33.

Musgrove, P. A. (1965). Lead: Grade–tonnage relation, *Mining Mag.* **112**, 249–51.

Nakanishi, H. (1990). Cellular-automaton model of earthquakes with deterministic dynamics, *Phys. Rev.* **A41**, 7086–9.

Nash, D. B. (1980a). Morphologic dating of degraded normal fault scarps, *J. Geol.* **88**, 353–60.

Nash, D. (1980b). Forms of bluffs degraded for different lengths of time in Emmet County, Michigan, USA, *Earth Surface Proc. Landforms* **5**, 331–45.

Newman, W. I. & Turcotte, D. L. (1990). Cascade model for fluvial geomorphology, *Geophys. J. Int.* **100**, 433–9.

Nicolis, C. & Nicolis, G. (1984). Is there a climatic attractor? *Nature* **311**, 529–32.

Nussbaum, J. & Ruina, A. (1987). A two degree-of-freedom earthquake model with static/dynamic friction, *Pure Appl. Geophys.* **124**, 629–56.

Pfeiffer, P. & Obert, M. (1989). Fractals: Basic concepts and terminology, in *The Fractal Approach to Heterogeneous Chemistry*, D. Avnir, ed., pp. 11–43, John Wiley, Chichester.

Plotnick, R. E. (1986). A fractal model for the distribution of stratigraphic hiatuses, *J. Geol.* **94**, 885–90.

Purcaru, G. & Berckhemer, H. (1982). Quantitative relations of seismic source parameters and a classification of earthquakes, *Tectonophys.* **84**, 57–128.

Rapp, R. H. (1989). The decay of the spectrum of the gravitational potential and the Earth, *Geophys. J. Int.* **99**, 449–55.

Rayleigh, Lord (1916). On convection currents in a horizontal layer of fluid when the higher temperature is on the underside, *Phil. Mag.* **32**, 529–46.

Reigber, C., Balmino, G., Muller, H., Bosch, W. & Moynot, B. (1985). GRIM gravity model improvement using LAGEOS (GRIM 3-Li), *J. Geophys. Res.* **90**, 9285–99.

Rice, J. R. & Tse, S. T. (1986). Dynamic motion of a single degree of freedom system following a rate and state dependent friction law, *J. Geophys. Res.* **91**, 521–30.

Richardson, L. F. (1961). The problem of contiguity: An appendix of statistics of deadly quarrels, *General Systems Yearbook* **6**, 139–87.

Rikitake, T. (1958). Oscillations of a system of disc dynamos, *Proc. Cambridge Phil. Soc.* **54**, 89–105.

Rosin, P. & Rammler, E. (1933). Laws governing the fineness of powdered coal, *J. Inst. Fuel* **7**, 29–36.

Ruina, A. (1983). Slip instability and state variability laws, *J. Geophys. Res.* **88**, 10359–70.

Rundle, J. B. (1988a). A physical model for earthquakes, 1. Fluctuation and interaction, *J. Geophys. Res.* **93**, 6237–54.

Rundle, J. B. (1988b). A physical model for earthquakes, 2. Application to Southern California, *J. Geophys. Res.* **93**, 6255–74.

Rundle, J. B. (1989). A physical model for earthquakes, 3. Thermo-dynamical approach and its relation to nonclassical theories of nucleation, *J. Geophys. Res.* **94**, 2839–55.

Rundle, J. B. & Jackson, D. D. (1977). Numerical simulation of earthquake sequences, *Seis. Soc. Am. Bull.* **67**, 1363–77.

Sadovskiy, M. A., Golubeva, T. V., Pisarenko, V. F. & Shnirman, M. G. (1985). Characteristic dimensions of rock and hierarchical properties of seismicity, *Phys. Solid Earth* **20**, 87–95.

Sammis, C. G. & Biegel, R. L. (1989). Fractals, fault-gauge, and friction, *Pure Appl. Geophys.* **131**, 255–271.

Sammis, C. G., Osborne, R. H., Anderson, J. L., Banerdt, M. & White, P. (1986). Self-similar cataclasis in the formation of fault gouge, *Pure Appl. Geophys.* **123**, 53–78.

Scheidegger, A. E. (1970). *Theoretical Geomorphology*, 2nd edition, Springer-Verlag, Berlin.

Schoutens, J. E. (1979). Empirical analysis of nuclear and high-explosive cratering and ejecta, in *Nuclear Geoplosics Sourcebook*, Vol. 55, part 2 section 4, Rep. DNA OIH-4-2 (Def. Nuclear Agency, Bethesda, MD).

Schreider, S. Yu (1990). Formal definition of premonitory seismic quiesence, *Phys. Earth Planet. Int.* **61**, 113–27.

Sieh, K. E. & Jahns, R. H. (1984). Holocene activity of the San Andreas fault at Wallace Creek, California, *Geol. Soc. Am. Bull.* **95**, 883–96.

Sieh, K., Stuiver, M. & Brillinger, D. (1989). A more precise chronology of earthquakes produced by the San Andreas fault in Southern California, *J. Geophys. Res.* **94**, 603–23.

Singh, S. K., Rodriguez, M. & Esteva, L. (1983). Statistics of small earthquakes and frequency of occurrence of large earthquakes along the Mexican subduction zone, *Seis. Soc. Am. Bull.* **73**, 1779–96.

Smalley, R. F., Chatelain, J. L., Turcotte, D. L. & Prevot, R. (1987). A fractal approach to the clustering of earthquakes: Applications to the seismicity of the New Hebrides, *Seis. Soc. Am. Bull.* **77**, 1368–81.

Smalley, R. F., Turcotte, D. L. & Sola, S. A. (1985). A renormalization group approach to the stick–slip behavior of faults, *J. Geophys. Res.* **90**, 1884–1900.

Smith, S. W. (1976). Determination of maximum earthquake magnitude, *Geophys. Res. Lett.* **3**, 351–4.

Sornette, A. & Sornette, D. (1989). Self-organized criticality and eathquakes, *Europhys. Lett.* **9**, 197–202.

Sornette, A. & Sornette, D. (1990). Earthquake rupture as a critical point: Consequences for telluric precursors, *Tectonophys.* **179**, 327–34.

Sparrow, C. (1982). *The Lorenz Equations: Bifurcations, Chaos, and Strange Attractors*, Springer-Verlag, New York, 269 pp.

Stauffer, D. (1985). *Introduction to Percolation Theory*, Taylor & Francis, London, 124 pp.

Stewart, C. A. & Turcotte, D. L. (1989). The route to chaos in thermal convection at infinite Prandtl number, 1. Some trajectories and bifurcations, *J. Geophys. Res.* **94**, 13707–17.

Stinchcombe, R. B. & Watson, B. P. (1976). Renormalization group approach for percolation conductivity, *J. Phys.* **C9**, 3221–47.

Takayasu, H. & Matsuzaki, M. (1988). Dynamical phase transition in threshold elements, *Phys. Lett.* **A131**, 244–7.

Taylor, S. R. (1964). Abundance of chemical elements in the continental crust: A new table, *Geochem. Cosmochim. Acta.* **28**, 1273–85.

Taylor, S. R. & McLennan, S. M. (1981). The composition and evolution of the continental crust: Rare earth element evidence from sedimentary rocks, *Phil. Trans. Roy. Soc. London* **301A**, 381–99.

Thompson, A. H., Katz, A. J. & Krohn, C. E. (1987). The microgeometry and transport properties of sedimentary rock, *Adv. Phys.* **36**, 625–94.

Todoeschuck, J. P., Jensen, O. G. & Labonte, S. (1990). Gaussian scaling noise model of seismic reflection sequences: Evidence from well logs, *Geophys.* **55**, 480–4.

Turcotte, D. L. (1986a). Fractals and fragmentation, *J. Geophys. Res.* **91**, 1921–6.

Turcotte, D. L. (1986b). A fractal model for crustal deformation, *Tectonophys.* **132**, 261–9.

Turcotte, D. L. (1986c). A fractal approach to the relationship between ore grade and tonnage, *Econ. Geol.* **81**, 1528–32.

Turcotte, D. L. (1987). A fractal interpretation of topography and geoid spectra on the earth, moon, Venus, and Mars, *J. Geophys. Res.* **92**, E597–E601.

Turcotte, D. L. (1989a). Fractals in geology and geophysics, *Pure Appl. Geophys.* **131**, 171–96.

Turcotte, D. L. (1989b). A fractal approach to probabilistic seismic hazard assessment, *Tectonophys.* **167**, 171–7.

Turcotte, D. L. & Schubert, (1982). *Geodynamics*, John Wiley and Sons, New York, 450 pp.

Verhulst, F. (1990). *Nonlinear Differential Equations and Dynamical Systems*, Springer-Verlag, Berlin, 277 pp.

Voss, R. F. (1985a). Random fractals: characterization and measurement, in *Scaling Phenomena in Disordered Systems*, R. Pynn & A. Skejeltorp, eds, pp. 1–11, Plenum Press, New York.

Voss, R. F. (1985b). Random fractal forgeries, in *Fundamental Algorithms for Computer Graphics*, NATO ASI Series, Vol. F17, R. A. Earnshaw, ed., pp. 805–35, Springer-Verlag, Berlin.

Voss, R. F. (1988). Fractals in nature: From characterization to simulation, in *The Science of Fractal Images*, H. O. Peitgen & D. Saupe, eds, pp. 21–70, Springer-Verlag, New York.

Walden, A. T. & Hosken, J. W. J. (1985). An investigation of the spectral properties of primary reflection coefficients, *Geophys. Prosp.* **33**, 400–35.

Wallace, R. E. (1977). Profiles and ages of young fault scarps, north-central Nevada, *Geol. Soc. Am. Bull.* **88**, 1267–81.

Walsh, J. J. & Watterson, J. (1988). Analysis of the relationship between displacements and dimensions of faults, *J. Struct. Geol.* **10**, 329–47.

Wilson, K. G. and Kogut, J. (1974). The renormalization group and the ε expansion, *Phys. Rev.* **C12**, 75–200.

Wyss, M. and Haberman, R. E. (1988). Precursory seismic quiescence, *Pure Appl. Geophys.* **126**, 319–32.

Youngs, R. R. and Coppersmith, K. J. (1985). Implications of fault slip rates and earthquake recurrence models to probabilistic seismic hazard estimates, *Seis. Soc. Am. Bull.* **75**, 939–64.

APPENDIX A

Glossary of terms

ATTRACTOR   A point in phase space toward which a time history evolves as transients die out.

BASIN OF ATTRACTION   Some dynamical systems have more than one fixed point. The region in phase space in which solutions approach a particular fixed point is known as the basin of attraction of that fixed point. The boundaries of a basin of attraction are often fractal.

BIFURCATION   A change in the dynamical behavior of a system when a parameter is varied.

CANTOR DUST   A fractal set generated by subdividing a line into parts.

CHAOS   Solutions to deterministic equations are chaotic if adjacent solutions diverge exponentially in phase space; this requires a positive Lyapunov exponent.

CLUSTER   A group of particles with nearest-neighbor links to other particles in the cluster.

DETERMINISTIC   A dynamical system whose equations and initial conditions are fully specified and are not stochastic or random.

DIFFERENCE EQUATION   An equation that relates a value of a function x_{n+1} to a previous value x_n. A difference equation generates a discrete set of values of the function x.

DIMENSION   The usual definition of dimension is the topological dimension. The dimension of a point is zero, of a line is one, of a square is two, of a cube is three. In this book we have introduced the concept of fractional (non-integer) dimensions, or fractals.

FEIGENBAUM NUMBER   The ratios of successive differences between period-doubling bifurcation parameters approach this number ($F = 4.669\,202$).

FIXED POINT   A point in phase space towards which a dynamical system approaches as transients die out.

FRACTAL   A power-law relation between the number of objects and their linear size. Also, a fractal is an object whose shape is independent of scale.

FRACTAL DIMENSION   The power value in the fractal relation.

HAUSDORFF DIMENSION   The power value in the power-law relationship between Fourier series coefficients and the corresponding wavelengths.

HOPF BIFURCATION   A bifurcation from a fixed point to a limit cycle.

LIMIT CYCLE   A periodic orbit in phase space towards which a dynmaic system approaches as transients die out.

LORENZ EQUATIONS   A set of three first-order differential equations derived from the equations governing thermal convection. Historically this was the first example of deterministic chaos.

LYAPUNOV EXPONENT   Solutions to deterministic equations are chaotic if adjacent solutions diverge exponentially in phase space; the exponent is known as the Lyapunov exponent. Solutions are chaotic if the Lyapunov exponent is positive.

MAP   A mathematical relation that translates one or more points into other points.

NODAL POINT   A fixed point towards which solutions evolve.

PERIOD DOUBLING   A sequence of periodic oscillations in which the period doubles as a parameter is varied.

PHASE SPACE   A coordinate space defined by the state variables of a dynamical system.

PITCHFORK BIFURCATION   A bifurcation in which the period doubles.

POINCARÉ SECTION   The sequence of points in phase space generated by the penetration of the evolving trajectory through a specified planar surface.

PROBABILITY   The likelihood that a particular event will occur. The probability that the next flip of a coin will be heads is 0.5.

RANDOM   A choice that is determined by pure chance. For example, a flip of a coin.

RENORMALIZATION   The transformation of a set of equations from one scale to another by a change of variables.

SADDLE POINT   A fixed point that attracts only a singular set of trajectories.

SCALE INVARIANCE   The phenomenon whereby an object appears identical at a variety of scales.

SELF-AFFINE   Under an affine transformation the different coordinates are scaled by different factors. If an object is scaled using an affine transformation the object is described as being self-affine.

SELF-SIMILARITY   A property of a set of points if their geometrical structure at one length scale is the same as at another length scale.

STRANGE ATTRACTOR   A fixed point in a phase space in which the orbits are chaotic.

Units and symbols

Table B1. *SI units*

Quantity	Unit	Symbol	Equivalent
Basic units			
Length	meter	m	
Time	second	s	
Mass	kilogram	kg	
Temperature	kelvin	K	
Electric current	ampere	A	
Derived units			
Force	newton	N	$\mathrm{kg\,m\,s^{-2}}$
Energy	joule	J	$\mathrm{kg\,m^2\,s^{-2}}$
Power	watt	W	$\mathrm{kg\,m^2\,s^{-3}}$
Pressure	pascal	Pa	$\mathrm{kg\,m^{-1}\,s^{-2}}$
Frequency	hertz	Hz	$\mathrm{s^{-1}}$
Charge	coulomb	C	$\mathrm{A\,s}$
Electric potential	volt	V	$\mathrm{kg\,m^2\,A^{-1}\,s^{-3}}$
Magnetic field	tesla	T	$\mathrm{kg\,A^{-1}\,s^{-2}}$
Multiples of 10			
10^{-3}	milli	m	
10^{-6}	micro	μ	
10^{-9}	nano	n	
10^{-12}	pico	p	
10^{3}	kilo	k	
10^{6}	mega	M	
10^{9}	giga	G	
10^{12}	tera	T	

Table B2. *Symbols*

Symbol	Quantity	Equation introduced	SI units
a	Parameter	(10.1)	
a_0	Radius of earth	(7.28)	m
$\dot{a}$	Frequency of earthquakes	(4.1)	s^{-1}
A	Area	(3.19)	m^2
b	Exponent	(3.1)	
	b-value for earthquakes	(4.1)	
c	Constant	(4.4)	
c_p	Specific heat at constant pressure	(12.4)	$J\,kg^{-1}\,K$
C	Constant	(2.1)	
	Concentration	(5.1)	
	Moment of inertia	(14.3)	$kg\,m^2$
d	Constant	(4.4)	
D	Fractal dimension	(2.1)	
E	Energy	(4.2)	J
f	Distribution function	(3.13)	
	Probability of fragmentation	(3.18)	
	Frequency	(7.13)	s^{-1}
F	Feigenbaum constant	(10.11)	
	Force	(11.1)	N
g	Acceleration due to gravity	(12.3)	$m\,s^{-2}$
G	Applied torque	(14.2)	N m
h	Elevation	(7.24)	m
	Layer thickness	(12.5)	m
H	Hausdorff measure	(7.1)	
	Fourier transform	(7.24)	m
I	Electrical current	(14.1)	A
J	Transport coefficient	(8.4)	$m^2\,s^{-1}$
k	Wave number	(7.31)	m^{-1}
	Spring constant	(9.12)	Nm^{-1}
	Thermal conductivity	(12.4)	$Wm^{-1}\,K^{-1}$
L	Length	(7.36)	m
	Self inductance	(14.3)	$Vs\,A^{-1}$
m	Mass	(3.1)	kg

Table B2. (*Cont.*)

Symbol	Quantity	Equation introduced	SI units
m	Earthquake magnitude	(4.1)	
M	Mass	(3.3)	kg
	Earthquake moment	(4.3)	J
	Mutual inductance	(14.1)	$V s A^{-1}$
n	Number	(2.1)	
Nu	Nusselt number	(12.42)	
$\dot{N}$	Number per unit time	(4.1)	s^{-1}
p	Pressure	(12.2)	Pa
	Probability	(6.1)	
P	Perimeter	(2.3)	m
Pr	Prandtl number	(12.13)	
r	Linear dimension	(2.1)	m
	Radius	(7.28)	m
	Ratio of Rayleigh numbers	(12.26)	
R	Rate coefficient	(8.23)	s^{-1}
	Resistance	(14.1)	$V A^{-1}$
Ra	Rayleigh number	(12.12)	
S	Power spectral density	(7.15)	
t	Time	(7.1)	s
T	Time interval	(7.5)	s
	Temperature	(12.4)	K
u	Horizontal velocity coordinate	(12.1)	ms^{-1}
v	Velocity	(4.11)	ms^{-1}
	Vertical velocity	(12.1)	ms^{-2}
V	Volume	(3.17)	m^3
	Variance	(3.16)	
x	Variable	(7.1)	
	Horizontal coordinate	(12.1)	m
X	Fourier transform	(7.13)	
y	Vertical coordinate	(12.1)	m
Y	Nondimensional position	(11.3)	
α	Constant	(4.5)	
	Stiffness parameter	(11.12)	
	Volume coefficient of thermal expansion	(12.8)	K^{-1}

Table B2. (*Cont.*)

Symbol	Quantity	Equation introduced	SI units
β	Power	(7.16)	
	Constant	(9.12)	
	Symmetry parameter	(11.12)	
$\dot{\beta}$	Constant	(4.6)	
γ	Constant	(8.25)	
Γ	Slope	(10.6)	
δ	Displacement across fault	(4.3)	m
ε	Parameter	(9.13)	
θ	Latitude	(7.28)	
	Polar coordinate	(9.23)	
	Temperature difference	(12.7)	K
κ	Thermal diffusivity	(12.11)	$m^2\,s^{-1}$
λ	Wavelength	(7.31)	m
	Lyapunov exponent	(10.19)	
μ	Parameter	(3.14)	
	Shear modulus	(4.3)	Pa
	Viscosity	(12.2)	Pa s
	Rikitake parameter	(14.8)	
ν	Power	(3.3)	
ρ	Density	(3.34)	$kg\,m^{-3}$
	Autocovariance	(8.12)	
	Polar coordinate	(9.28)	
σ	Parameter	(3.14)	
	Standard deviation	(7.7)	
τ	Time interval	(4.22)	s
	Nondimensional time	(11.3)	
ϕ	Porosity	(3.33)	
	Enrichment factor	(5.1)	
	Longitude	(7.28)	
	Friction parameter	(11.3)	
ψ	Stream function	(12.6)	$m^2\,s^{-1}$
Ω	Angular velocity	(14.1)	s^{-1}

Answers to selected problems

2.1 $N_3 = 8$, $r_3 = 1/27$, $N_4 = 16$, $r_4 = 1/81$.

2.2 $N_3 = 27$, $r_3 = 1/125$, $N_4 = 81$, $r_4 = 1/625$.

2.3   (b) $N_1 = 2$, $N_2 = 4$, $N_3 = 8$, $r_1 = 1/5$, $r_2 = 1/25$, $r_3 = 125$.
     (c) $D = 0.4307$.

2.4   (b) $N_1 = 3$, $N_2 = 9$, $N_3 = 27$, $r_1 = 1/7$, $r_2 = 1/49$, $r_3 = 1/343$.
     (c) $D = 0.5646$.

2.5   (b) $N_1 = 4$, $N_2 = 16$, $N_3 = 64$, $r_1 = 1/7$, $r_2 = 1/49$, $r_3 = 1/343$.
     (c) $D = 0.7124$.

2.6 $N_3 = 512$, $N_4 = 4096$, $r_3 = 1/27$, $r_4 = 1/81$.

2.7   (b) $N_1 = 2$, $N_2 = 4$, $N_3 = 8$, $r_1 = 1/2$, $r_2 = 1/4$, $r_3 = 1/8$.
     (c) $D = 1$.

2.8   (b) $N_1 = 12$, $N_2 = 144$, $N_3 = 1728$, $r_1 = 1/4$, $r_2 = 1/16$, $r_3 = 1/64$.
     (c) $D = 1.79$.

2.9   (b) $N_1 = 24$, $N_2 = 576$, $N_3 = 13\,824$, $r_1 = 1/5$, $r_2 = 1/25$, $r_3 = 1/125$.
     (c) $D = 1.9746$.

2.10 (b) $N_1 = 17$, $N_2 = 289$, $N_3 = 4913$, $r_1 = 1/5$, $r_2 = 1/25$, $r_3 = 1/125$.
     (c) $D = 1.76$.

2.11 $D = 2.975$.

2.13 (b) $A_1 = 1/2$, $A_2 = 12/2$, $A_3 = 120/2$.   (c) No.

2.14 (a) $N_1 = 1$, $N_2 = 8$, $N_3 = 64$, $r_1 = 1/3$, $r_2 = 1/9$, $r_3 = 1/27$.
     (b) Yes.   (c) $D = 1.89$.

3.1 $\bar{r} = vr_0/(v + 1)$, $V = v(v + 3)(v - 1)r_0^2/(v + 2)(v + 1)^2$.

3.4 $N_1 = 4$, $N_2 = 16$, $N_3 = 64$, $D = 2$.

3.5 $\phi_1 = 0.0385$, $\rho_1 = 0.9627\rho_0$, $\phi_2 = 0.0727$, $\rho_2 = 0.9273\rho_0$, $D = 2.9656$.

3.6 $\phi_1 = 0.2857$, $\rho_1 = 0.7778\rho_0$, $\phi_2 = 0.3951$, $\rho_2 = 0.6049\rho_0$, $D = 2.7712$.

4.1 $E_s = 2.1 \times 10^{15}$ J, $M = 4 \times 10^{19}$ J, $A = 530\,\text{km}^2$, $\delta_e = 2.5\,\text{m}$.

4.2 $E_s = 7.1 \times 10^{13}$ J, $M = 1.3 \times 10^{18}$ J, $A = 54\,\text{km}^2$, $\delta_e = 0.8\,\text{m}$.

4.3   100.

4.4   11.9 yr.

4.5 10^3 yr.

4.6   (a) $\dot{a} = 2 \times 10^6\,\text{yr}^{-1}$. (b) $\tau_e = 158$ yr.

4.7   (a) $\dot{a} = 7 \times 10^4\,\text{yr}^{-1}$. (b) $\tau_e = 22$ yr.

4.8   400 yr.

4.9   1600 yr.

4.10 1000 yr.

5.1   1.57.

5.2   1.26.

5.3   (a) $(8 - \phi_8)C_0/7$. (b) $(8 - \phi_8)\phi_8 C_0/7$. (c) $1 < \phi_8 < 8$.
      (d) $0 < D < 3$.

5.4 3.18×10^8 kg.

5.5 8.37×10^{10} kg.

5.6 8.18×10^8 kg.

6.1   1, 1/3, 1/9, 1/27.

6.2   1, 1, 1, 1.

6.3   1, 3/5, 9/25.

6.4   1, 4/7, 16/49, 64/343.

6.5   1, 3/7, 9/49, 27/343.

6.6   1, 1/3, 1/9, 1/27.

6.7   1, 1/2, 1/4, 1/8.

6.8   1, 17/25, 289/625.

6.9   1, 6/8, 36/64, 216/512.

6.10 1, 26/27, 676/729.

9.1 $y = c_1 e^{at}$, $x = c_2 e^{ft}$.

9.3 $x_1 = c_1 e^{\mu t}$ (diverges), $x_2 = c_2 e^{-2\mu t}$ (converges), $x_3 = c_3 e^{-2\mu t}$ (converges).

9.5 $x = 1$ (stable), $x = -1$ (unstable).

10.1 $x_1 = 0.125$, $x_2 = 0.0547$, $x_3 = 0.0258$, $x_4 = 0.0126$, $x_f = 0$.

10.2 $x_1 = 0.16875$, $x_2 = 0.12625$, $x_3 = 0.09928$, $x_4 = 0.08048$, $x_f = 0$.

10.3 $x_1 = 0.32$, $x_2 = 0.4352$, $x_3 = 0.4916$, $x_4 = 0.4999$, $x_f = 0.5$.

10.4 $x_1 = 0.525$, $x_2 = 0.6234$, $x_3 = 0.5869$, $x_4 = 0.6061$, $x_f = 0.6$.

10.5 $x_{f1} = 0.513045$, $x_{f2} = 0.799455$.

10.6 $x_{f1} = 0.45197$, $x_{f2} = 0.84216$.

10.7 $x_{fmax} = 0.925$, $x_{fmin} = 0.2567$.

10.8 $x_{fmax} = 0.95$, $x_{fmin} = 0.1805$.

10.9 $x_0 = 0.2298$, $x_1 = 0.7081$, $x_2 = 0.8268$, $x_3 = 0.5728$, $x_4 = 0.9788$.

10.10 $x_0 = 0.1070$, $x_1 = 0.3824$, $x_2 = 0.9447$, $x_3 = 0.2091$, $x_4 = 0.6615$.

11.1 (a) $Y = 1$. (b) $Y = 1.333$.

11.2 (a) $Y = 1$. (b) $Y = -0.3333$.

11.3 $V = \left[\left(1 - \frac{1}{\phi}\right)^2 - \left(Y - \frac{1}{\phi}\right)^2 \right]^{1/2}$.

11.4 (a) $Y_1 = 1$, $Y_2 = 0.5$. (b) $Y_1 = 0$, $Y_2 = 0.5$. (c) $Y_1 = 0.5$, $Y_2 = 1$.
(d) $Y_1 = 0.5$, $Y_2 = 0$.

11.5 (a) $Y_1 = 1$, $Y_2 = 0.75$. (b) $Y_1 = 0.5$, $Y_2 = 0.75$. (c) $Y_1 = 0.75$, $Y_2 = 1$.
(d) $Y_1 = 0.75$, $Y_2 = 0.5$.

13.1 -6.16, -1.714.
13.2 14.21, 7.20.

15.1 12, 144.
15.2 $n_e = n^{D/3}$.
15.5 34.
15.6 33.
15.9 0.094.

16.1 (a) 1, 5, 6, 6. (b) 6, 1; 5, 0; 2, 1; 4, 1; 8, 1.
16.2 (a) 7, 8, 2, 4. (b) 3, 2; 6, 1; 5, 0; 9, 2; 1, 2; 2, 1.
16.3 (a) 5, 6. (b) 7, 2; 4, 1; 8, 1; 5, 0; 6, 1.

Index

absolute value, 77
acoustic impedance, 84
acoustic well log, 84
aftershock statistics, 192
alluvial fan, 14, 97
amplitude, 77
analysis
    screen, 21
    sieve, 21
angle of repose, 186
Appalachian earthquakes, 43
Arizona topography, 92
Armenian earthquake, 92
asperities, 179
atmosphere, 114
attractor, 206
    strange, 146, 208
autocovariance function, 99

basin
    drainage, 96
    of attraction, 206
bathymetry, 79, 85
bifurcation, 110, 207
    diagram, 110, 120, 131
    flip, 117
    Hopf, 112, 145, 207
    pitchfork, 111, 131, 142, 207
    turning point, 110
bin analysis, 27
binomial distribution, 54
block model, 4, 125, 190
boundary layer, thermal, 149, 151
Boussinesq approximation, 138,
    152
box-counting method, 15
    for time series, 76
Brownian noise, 3, 74
    fractional, 75, 78, 79
buoyancy force, 138, 139
b-value, 2, 36

cable, stranded, 179
Cantor dust, 8, 206
Cantor set, 8, 65
carpet, Sierpinski, 9, 70, 171
cascade, geomorphology, 99
cell, 25, 53, 172
cellular-automata model, 5, 187
center, 107
chaos, 3, 4, 206
    deterministic, 104, 114, 137, 149
    route to, 120
    windows of, 120, 134
characteristic earthquake, 44
Charleston earthquake, 43
climate, 85
cluster, 206
    percolation, 171
clustering, 65
    fractal, 65
    seismicity, 68, 192
coal, 23
coastline, length of, 12
comminution, 29
complex quantity, 77
conduction, heat, 138
conductivity, electrical, 175
continental drift, 192
continuity equation, 137
convection
    heat, 138
    mantle, 151
    thermal, 3, 137
copper, 60
core, 92
correlation, 74
creep, thermally activated, 151
critical phenomena, 5, 186
critical point, 174, 186
critical probability, 170
critical Rayleigh number, 141
critical state, 41

criticality, self-organized, 5, 186
crustal deformation, 35
Culling model, 97
curdling, 8
Curie temperature, 192
cycle, limit, 108, 117, 207

damping
    linear, 106
    nonlinear, 106
Darcy's law, 170
deformation, crustal, 35
delta, prograding, 97
density
    power spectral, 77, 86
    soil, 32
deposits, ore, 52
deterministic, 206
deterministic chaos, 104, 114, 137, 149
deterministic fractal, 12
deviation, standard, 76
devil's staircase, 11
difference equation, 206
diffusion equation, 97
dimension, 206
    Euclidean, 6
    fractal, 6, 207
        of earthquakes, 37
        of time series, 76
discrete Fourier transform, 78
    inverse, 79
disk dynamo, 163
displacement
    earthquake, 36
    fault, 48
distributed seismicity
    global, 38
    in southern California, 40
distribution
    binomial, 54
    fractal
        of earthquakes, 37
        of faults, 44
    Gaussian, 75
    log-normal, 22, 52
    mass, 22
    mean of, 22
    number, 22
    Poisson, 67
    power-law, 2, 9, 20, 52
    Rosin and Rammler, 21
    variance, 23
    Weibull, 21, 180
distribution function, 22
doubling, period, 117, 119, 131, 207
drainage basin, 96
drainage patterns, 95
drift, continental, 192
dust, Cantor, 8, 206
dynamic friction, 126

dynamical systems, 104
dynamics, population, 104
dynamo, 162
    disk, 163
    Rikitake, 4, 163
    self-excited, 162

earthquake
    Armenian, 192
    characteristic, 44
    Charleston, 43
    Kern County, 35
    Loma Prieta, 35, 192
    San Fernando, 35
earthquake displacement, 37
earthquake energy, 36
earthquake frequency, 68
earthquake magnitude, 36
earthquake moment prediction, 36,
        192
earthquake rupture area, 38
earthquakes, 4, 17, 125, 179, 188
    Appalachian, 43
    clustering of, 68, 192
    fractal dimension of, 37
    fractal distribution of, 37
    frequency–magnitude statistics for, 2, 36,
        188
    frequency–size statistics for, 17
    intervals between, 48
    Mississippi Valley, 42
    New England, 43
    Parkfield, 44, 134
economic ore deposits, 52
elastic rebound, 125
electrical conductivity, 175
electrical conductivity well log, 84
element, 25, 53, 172
energy balance
    earthquake, 36
energy balance equation, 138
enrichment factor, 53
erosion, 14, 95
    characteristic time for, 97
erosional processes, 95
eruptions, volcanic, 20, 49
Euclidean dimension, 6
expansion, thermal, 137
    coefficient, 139
explosion, nuclear, 23
explosive processes, 20

fan, alluvial, 14, 97
fault
    rupture, 179
    San Andreas, 35, 40
fault displacement, 48
fault gouge, 30
fault scarps, 97
faults, 20

faults (*cont.*)
  fractal distribution of, 44
  transform, 35
Feigenbaum constant, 119, 206
Feigenbaum relation, 119
field, magnetic, 162
fields, oil, 62, 85
filtering, 79
fixed point, 105, 115, 173, 206
  stability of, 105, 173
flip bifurcation, 117
floods, 84, 102
fluid layer, 137
force balance equations, 138
  buoyancy, 138, 139
  inertial, 138
  pressure, 138
  viscous, 138
Fourier coefficients, 78
Fourier transform, 77
  discrete, 78
  inverse, 77
  two-dimensional, 88
fractal, 1, 6, 207
  clustering, 65
  deterministic, 12
  island, 12
  self-affine, 73
  self-similar, 74
  statistical, 12
fractal dimension, 6, 207
  of earthquakes, 37
  of time series, 76
fractal distribution, 6, 15
  of earthquakes, 37
  of faults, 44
fractal tree, 2, 179
fractional Brownian noise, 75, 78, 79
fractures, 20
fragmentation, 20, 176
fragmentation probability, 26, 178
fragments, rock, 15
  size distribution of, 20
free surface, 141
frequency-magnitude statistics
  for earthquakes, 2, 36, 188
frequency-size statistics
  for earthquakes, 17
  for fragments, 20
  for islands, 17
friction, 125, 179
  dynamic, 126
  static, 126

Gaussian distribution, 75
Gaussian white noise, 78
geoid, 87
geomorphology, 14, 95
geomorphology cascade, 99
glaciation, 102

global seismicity, 38
gouge, fault, 24, 30
  synthetic, 30
grade, ore, 52
group, renormalization, 5, 169
Gutenberg–Richter relation, 2, 36, 188

harmonic oscillator, 106
harmonics, spherical, 85
Hausdorff measure, 74, 75, 207
hazards, seismic, 41
heat conduction, 138
heat conduction equation, 4, 14, 114
heat convection, 138
hiatuses, stratigraphic, 85
Hopf bifurcation, 112, 145, 207
hot spots, 151
hydrology, 170
hydrothermal circulations, 52

impacts, 20, 23
impedance, acoustic, 84
impermeable element, 170
inductance, 163
inertial force, 138
intervals between earthquakes, 48
invariance, scale, 1, 8, 207
inverse Fourier transform, 77
  discrete, 79
island, fractal, 12
island, Koch, 11
islands, frequency-size statistics for, 17
isotope ratios, oxygen, 85
isotropic, 73
iterative map, 115

Kern County earthquake, 35
Koch island, 11
Korcak relation, 17

landscapes, 95
Laplace equation, 4, 114
layer, fluid, 137
Legendre function, 87
length of coastline, 12
length of perimeter, 12
limit cycle, 108, 117, 207
linear dampling, 106
linearization, 105, 140
logarithmic spiral, 110
logistic equation, 104
logistic map, 4, 114, 173
log-normal distribution, 22, 52
logs, well, 83
Loma Prieta earthquake, 35, 192
Lorenz equations, 3, 137, 143, 152, 207
Lyapunov exponent, 4, 121, 134, 207

magnetic field, 162
magnetic field polarity, 192

magnetic field reversals, 4, 192
magnetic field surveys, 85
magnetism, natural remanent, 192
magnitude
    body wave, 36
    earthquake, 36
    local, 36
    Richter scale for, 36
    surface wave, 36
mantle convection, 151
mantle plume, 151
map, 207
    iterative, 115
    logistic, 4, 114, 173
mass conservation, 97
mass distribution, 22
mean, 22, 75
measure, Hausdorff, 74, 75, 207
medium, porous, 30, 170
Menger sponge, 10, 30, 70, 175
mercury, 58
Mississippi Valley earthquakes, 42
model
    cellular-automata, 5, 187
    slider-block, 4, 125, 190
moment, 76
    of earthquake, 36
Monte Carlo approach, 171
motion, equation of, 106
mountain building, 35

Nankai trough, 134
natural remanent magnetism, 192
Navier–Stokes equations, 114
nearest-neighbor model, 187
neutron activation well log, 84
New England earthquakes, 43
node, 109, 207
    stable, 109
    unstable, 109
noise
    Brownian, 3, 74
    fractional Brownian, 75, 78, 79
    Gaussian white, 78
    white, 75
nondimensional parameters, 107, 139
nondimensional variables, 104, 139
nonlinear damping, 106
nonlinear equations, 104, 114
no-slip, 141
nuclear explosion, 23
number distribution, 22
Nusselt number, 145

ocean ridge, 35, 151
ocean trench, 35, 151
oceans, 114
oil fields, 62, 85
orbit, periodic, 129
order, 8

of river, 96
ore deposits, 52
ore grade, 52
ore reserves, 52
ore tonnage, 52
Oregon topography, 81, 91
oscillator, harmonic, 106
oscillator system, 106
oxygen isotope ratios, 85

paleomagnetism, 192
Parkfield earthquakes, 44, 134
pattern recognition, 192
patterns, drainage, 95
percolation, 173
percolation cluster, 171
percolation threshold, 171
perimeter, length of, 12
period doubling, 117, 119, 131, 207
periodic orbit, 129
periogram method, 81
permeable, 170
permeability, 30, 175
petroleum engineering, 170
petroleum reserves, 62
phase, 77
phase change, 169
phase plane, 107
phase portrait, 146
phase space, 107, 207
phase trajectory, 107
pile, sand, 186
pitchfork bifurcation, 111, 131, 142, 207
plane, phase, 107
plate tectonics, 35, 42, 151
plume, mantle, 151
Poincaré section, 207
point
    critical, 174, 186
    fixed, 105, 115, 173, 206
    saddle, 109, 207
    singular, 108
    stable fixed, 105, 173
Poisson distribution, 67
polarity, magnetic, 192
population dynamics, 104
porosity, 30, 84, 175
porosity logs, 84, 85
porosity structure, 85
porous medium, 30, 170
portrait, phase, 146
power, 77
power-law distribution, 2, 9, 20, 52
power spectral density, 77, 86
Prandtl number, 139, 152
prediction of earthquakes, 192
pressure force, 138
probability, 65, 207
    critical, 170
    fragmentation, 26, 178

prograding delta, 97
pull, trench, 42
push, ridge, 42

quadratic Weibull distribution, 21, 180
quiescence, seismic, 192

random choice, 207
random walk, 3, 74
Rayleigh number, 139
  critical, 141
rebound, elastic, 125
recognition, pattern, 192
recursive relation, 115, 173
reflection coefficient, 84
remanent magnetism, 192
renormalization, 25, 55, 207
renormalization group method, 5, 169
repose, angle of, 186
reserves
  ore, 52
  petroleum, 62
reversals, magnetic field, 4, 192
Richter magnitude scale, 36
ridge, ocean, 35, 151
ridge push, 42
Rikitake dynamo, 4, 163
river network, 2
  order of, 96
rock fragments, 15
rock surfaces, 84
Rosin and Rammler distribution, 21
roughness, 83
route to chaos, 120
ruler method, 14
rupture area, earthquake, 37

saddle point, 109, 207
San Andreas fault, 35, 40
San Fernando earthquake, 35
sand pile, 186
scale invariance, 1, 8, 207
scarps
  earthquake, 97
  shoreline, 97
screen analysis, 21
section, Poincaré, 207
seismic hazards, 41
seismic network, 35
seismic quiescience, 102
seismicity, 35
  clustering of, 68, 192
  global, 38
  in southern California, 40
self-affine, 207
self-affine fractal, 73
self-organized criticality, 5, 186
self-similar, 1, 208
self-similar fractal, 74
series, time, 74

set, Cantor, 8, 65
shoreline scarps, 97
Sierpinski carpet, 9, 70, 171
sieve analysis, 21
singular point, 108
slider-block model, 4, 125, 190
soil density, 32
sonic well log, 84
southern California seismicity, 40
space, phase, 107, 207
spectral analysis, 3
spectral density, power, 77, 86
spectral techniques, 74
spherical harmonics, 85
spiral, logarithmic, 110
sponge, Menger, 10, 30, 175
spreading centers, 35
spring–mass oscillator, 106
stability analysis, 105, 140
stable fixed point, 105, 173
stable node, 109
staircase, devil's, 11
standard deviation, 76
static friction, 126
statistical fractal, 12
stick–slip behavior, 125
stiffness parameter, 128
storms, 84, 102
stranded cable, 179
strange attractor, 146, 208
stratigraphic hiatuses, 85
stream function, 138
subduction zones, 35, 151
surfaces, rock, 84
synthetic fault gouge, 30

tectonics, 14, 35
  plate, 35, 42, 151
temperature, 79,
  Curie, 192
tephra, 49
terraces, wave-cut, 11
thermal boundary layer, 149, 151
thermal convection, 3, 137
thermal expansion, 137
thermally activated creep, 151
threshold, percolation, 171
time series, 74
  box counting method for, 76
  fractal dimension of, 76
tonnage, ore, 52
topography, 2, 14, 73, 79, 85
  of Arizona, 92
  of Oregon, 81, 91
  global, 86
trajectory, phase, 107
transform
  discrete Fourier, 78
  Fourier, 77
  inverse discrete Fourier, 79

inverse Fourier, 77
two-dimensional Fourier, 88
transform faults, 35
tree, fractal, 2, 179
trench, ocean, 35, 151
trench pull, 42
triadic Koch island, 11
truncations, 153
turbulence, 3, 114
turning point bifurcation, 110
two-dimensional Fourier transform, 88

unstable fixed point, 105, 173
unstable node, 109
uranium, 61

van der Pol equation, 106, 112
variance, 23, 76, 86

viscous force, 138
volcanic edifices, 14
volcanic eruptions, 20, 49
volume of, 49

walk, random, 3, 74
wave-cut terraces, 11
wave equation, 4, 114
Weibull distribution, 21
quadratic, 21, 180
well logs, 83
acoustic, 84
electrical conductivity, 84
neutron activation, 84
porosity, 85
sonic, 84
white noise, 75
Gaussian, 78
windows of chaos, 120, 134